Aircraft Performance and Design

An Introduction to Principles and Practice

Ambar K. Mitra

Library of Congress
Copyright ©
Aircraft Performance and Design
April 1, 2019
ISBN 9 780999 746615

ambar.k.mitra@gmail.com

Foreword

For thousands of years –if not longer—humans have gazed toward the heavens. The pull in our souls to be "up there," soaring through the skies and beyond, has been perhaps, just as strong as the gravitational force that keeps us earthbound. Thanks to early efforts from the inquisitive minds of people like Leonardo DaVinci; Orville and Wilbur Wright; and early parachutist Georgia "Tiny" Broadwick, today we bask in the knowledge of having "conquered the skies" for over a century. Possessing key understanding in all areas for what it takes to soar safely and efficiently in atmospheric flight, the world continues – through new technology development and advancements-- to push the boundaries more than ever before.

Students undertaking a well-designed curriculum in aerospace engineering, as with all engineering disciplines, must start with the basics. A strong and well-understood foundation of principles cemented into their individual "mental tool-kits," is a must for a successful future in the aviation and associated industries. In *Aircraft Performance and Design,* atmospheric flight is introduced in a unique way to sophomore-level students, with the aim of exposing and exploring these principles in enough rigor, but without the sometimes overwhelmingly complex, mathematics.

Using appropriate level physics and calculus, *Aircraft Performance and Design* allows students to quickly and successfully gain a basic and usable understanding behind the key aircraft fundamentals of aerodynamics, performance, propulsion, and stability and control. For students digesting this curriculum, it should lead to a solid grasp of –and quite possibly a fascination and excitement for—all vehicles that fly.

Focusing on the entire gamut of an airplane's performance envelope, including cruise, takeoff, landing, and turns; with discussions on static stability, propeller- and jet engine-driven aircraft, the text is an excellent starting point for future, more in-depth learning.

As a United States Astronaut, I spent considerable time in both propeller and jet engine-driven aircraft, often times finding myself lacking an in-depth understanding of some of the key principles involved. A text like *Aircraft, Performance and Design* would have been a distinct advantage when applied as a useful tool in my astronaut tool belt. Fasten your seatbelts and prepare for takeoff!

Clayton C. Anderson
U.S. Astronaut, Retired

Preface

This book is an introduction to the principles of flight, such as, aerodynamics, propulsion, and static stability. Furthermore, the book introduces the analysis of various aspects of an airplane's mission, such as, takeoff, climb, cruise, descent and landing, and bank turn.

Readers with first year, college level proficiency in calculus and physics will fully appreciate the contents.

The purpose is to keep the students, in an Aeronautical Engineering program, anchored to the basic figures of merit and preliminary design of an airplane without overwhelming mathematical analysis. Even the graduating seniors will find this book as a practical guide for their Capstone Design projects.

The author would like to thank all his students for their inquiries that made him learn and all his colleagues for their teaching and insight.

The author's wife, Jayati Mitra has been the inspiration behind this book and the book would have remained incomplete without her encouragement and patience.

Table of Contents

1 Introduction

1.1 Audience

This book addresses the fundamentals of aircraft performance and design and the required background of readers is sophomore level physics and calculus. The purpose is to excite the sophomores, in an aerospace program, about airplanes without overwhelming them with differential equations and complex calculus. People with a fascination of airplanes will find many of their 'how, why, and what' questions answered in this book. Even seniors will find the concepts and practices useful for their capstone design projects.

1.2 Analysis, Synthesis, and Design

In this book, we will focus on three essential abilities for design.

- Analysis: Given the operating system, the airplane, and its operating environment, the altitude, determine the performance characteristics.
  - Given the airplane configuration: takeoff weight, engine thrust, wing area, drag, wing geometry, atmospheric condition (density)
  - Determine: cruise speed in level flight, service ceiling, climb rate, takeoff distance, stall speed
- Synthesis: Given the operating environment and required, performance characteristics, determine the operating system.
  - Given the altitude and landing distance
  - Determine: wing geometry, angle of attack, amount of flap
- Design: Determine the system that provides acceptable performance characteristics through repeated analysis and synthesis.
  - Given range, altitude, and payload
  - Determine amount of fuel consumption

1.3 Structure of an Aircraft Company

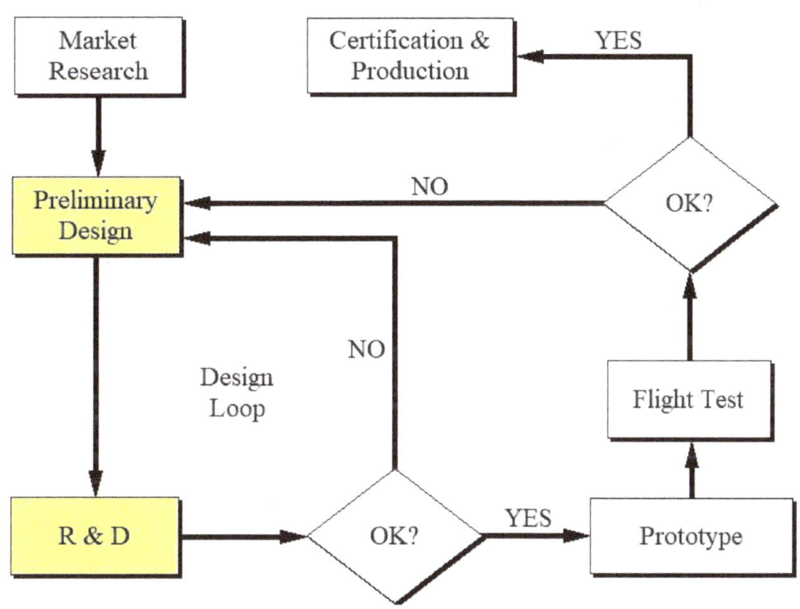

The path from market research to production contains two loops – one is the design loop and the other is the prototype-testing loop. Readers will find a rudimentary overview of preliminary design in Chapter 10 and of R&D in the remaining Chapters.

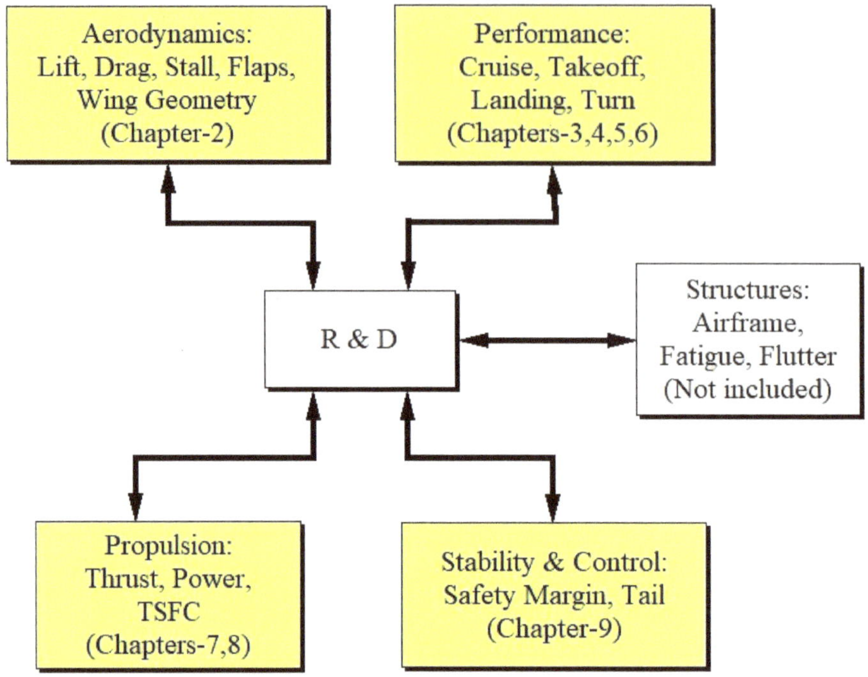

1.4 Spreadsheet

The performance and design calculations are often long and sometimes requiring the solution of non-linear equations. As shown in Appendices B through F, the Microsoft Excel utility is a convenient way of doing these calculations without human error.

1.5 Standard Atmosphere

The properties for US Standard Atmosphere are in Appendix G. For many calculations, it is convenient to have a formula connecting density and altitude. To obtain this formula we plot density against altitude and fit an equation to this plot.

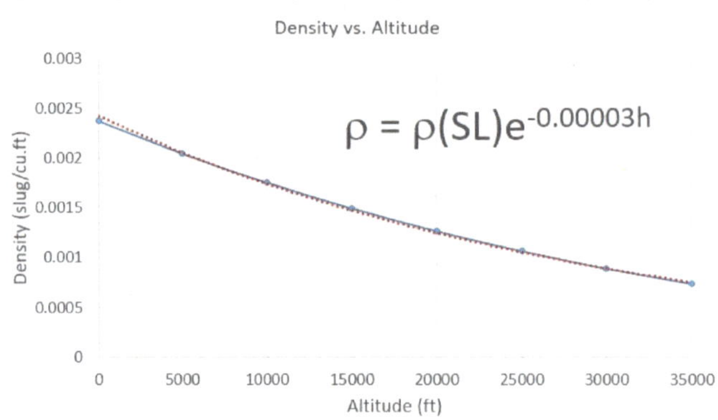

$$\rho = \rho_{SL} e^{-\frac{h}{33333}} \tag{1.1}$$

2 Incompressible Aerodynamics

2.1 Introduction

In this chapter, we will condense the contents of a yearlong aerodynamics course into about fifty pages. The purpose is to present the essential physics of flight and the equations of practice for lift and drag calculation, without overwhelming derivations.

Equations of airflow are non-linear and the geometry of an aircraft is complex with several components, such as the fuselage, the wing, and the tail. Detail calculation of lift and drag requires computer programs that are either commercial, such as Fluent and StarCCM, or legacy codes within an organization. The execution time for such programs is from hours to days.

At the beginning stages of design, we usually have many options to consider. Running the computer programs for all the options is prohibitive due to large expenses of time and effort. By using the equations of practice in this chapter, we can identify the good options and discard the bad ones. We can then do time consuming calculations and perform wind tunnel testing for the good options only.

2.2 Continuum Hypothesis

Aerodynamics is the study of dynamics of air. When we look at air through a microscope with sufficient magnification, we can see the nitrogen, oxygen, and other molecules. These molecules collide among themselves and with the aircraft.

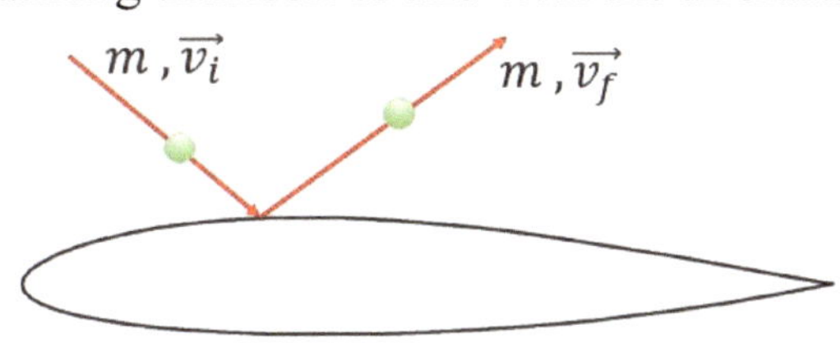

Figure 2.1: Collision of one molecule with the wing.

The impulse-momentum principle for the collision of one molecule of mass m with the wing is

$$m\vec{v_f} - m\vec{v_i} = \vec{F}\Delta t$$

Force $\vec{F}$ is on the molecule from the wing, $-\vec{F}$ is the force on the wing from the molecule, and the duration of the collision is Δt . Net force on a wing, such as lift and drag, is the sum of all forces from the infinitely many molecular collisions. However, calculating lift and drag from molecular collisions is impractical because we need to follow infinitely many molecules by tracking their position and velocity, and how the velocity changes due to collisions among the molecules and the wing.

Instead of this molecular approach, aerodynamicists view the motion of air in a macroscopic scale and the mathematics of motion does not include the molecular structure of air. In this macroscopic scale, the molecules or the voids between the molecules are not visible and air appears as a continuum.

A particle of air is sufficiently small so that all parts of the particle have the same density, velocity, temperature, momentum, etc., but not too small that the molecules are visible.

2.3 Lagrangian and Eulerian Descriptions

In the Lagrangian description, we number the fluid particle as *1, 2, 3 ...* and their properties, such as position, velocity *(m/s, ft/s)*, pressure *(lb/ft², N/ft²)*, are functions of time.

$$\overrightarrow{r_1}(t), \overrightarrow{r_2}(t), \overrightarrow{r_3}(t)$$

$$\overrightarrow{v_1}(t), \overrightarrow{v_2}(t), \overrightarrow{v_3}(t)$$

$$\overrightarrow{p_1}(t), \overrightarrow{p_2}(t), \overrightarrow{p_3}(t)$$

However, engineers are interested in the properties at a particular location, such as the velocity or pressure at points *P* and *Q* on a wing, and not the properties of individual particles.

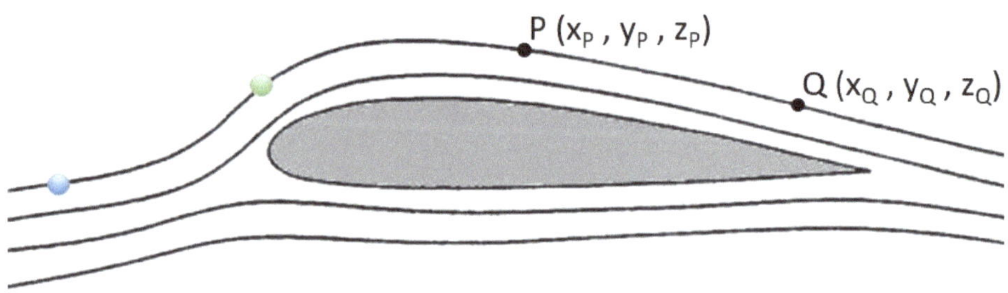

Figure 2.2: Lagrangian and Eulerian descriptions.

In Eulerian description, pressure, and velocity at *P* and *Q* are

$$p(x_P, y_P, z_P, t) \;;\; \vec{v}(x_P, y_P, z_P, t)$$
$$p(x_Q, y_Q, z_Q, t) \;;\; \vec{v}(x_Q, y_Q, z_Q, t)$$

When the green or the blue particle arrives at *P* or *Q*, the particle acquires the Eulerian density and velocity of the point. If the green particle occupies point *P* at time t_1 and point *Q* at time t_2, then the relationship between Lagrangian and Eulerian descriptions becomes

$$\vec{v}(x_P, y_P, z_P, t_1) = \overrightarrow{v_{Green}}(t_1) \;;\; \vec{v}(x_Q, y_Q, z_Q, t_2) = \overrightarrow{v_{Green}}(t_2)$$

If the blue particle occupies point *P* at time t_3 and point *Q* at time t_4, then the relationship between Lagrangian and Eulerian descriptions becomes

$$\vec{v}(x_P, y_P, z_P, t_3) = \overrightarrow{v_{Blue}}(t_3)\ ;\ \ \vec{v}(x_Q, y_Q, z_Q, t_3) = \overrightarrow{v_{Blue}}(t_4)$$

These relationships are of theoretical interest only. Engineers do not use the Lagrangian description because the Eulerian description is a better alternative compared to the Lagrangian description.

2.4 Airspeed and Ground Speed

Let $\overrightarrow{v_{W/G}}$ be the wind velocity with respect to the ground, $\overrightarrow{v_{A/W}}$, the airspeed, is the velocity of the airplane with respect to the surrounding mass of air, and $\overrightarrow{v_{A/G}}$, the ground speed is the velocity of the airplane with respect to the ground. The relationship among these three velocities is

$$\overrightarrow{v_{A/G}} = \overrightarrow{v_{A/W}} + \overrightarrow{v_{W/G}} \tag{2.1}$$

In the rest of this book, velocity will always refer to the airspeed because the pilot can view this on the console. There is no gauge in the cockpit to display the wind velocity and ground speed.

Pilots receive the wind speed information from the Aviation Weather Center.

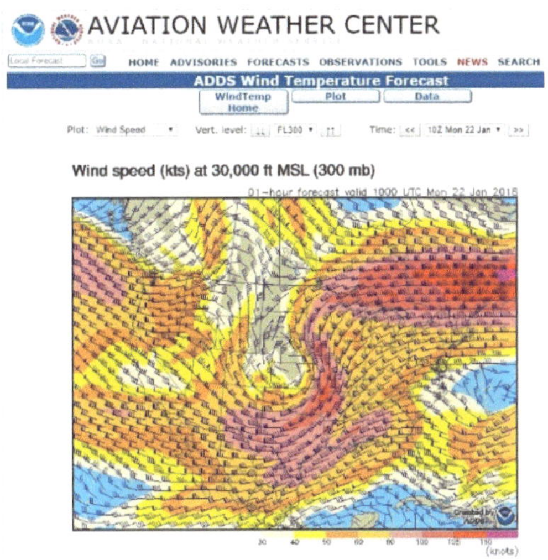

Figure 2.3: Wind speed.

An airspeed indicator and a compass, on pilot's console, display the magnitude and the direction of the airspeed.

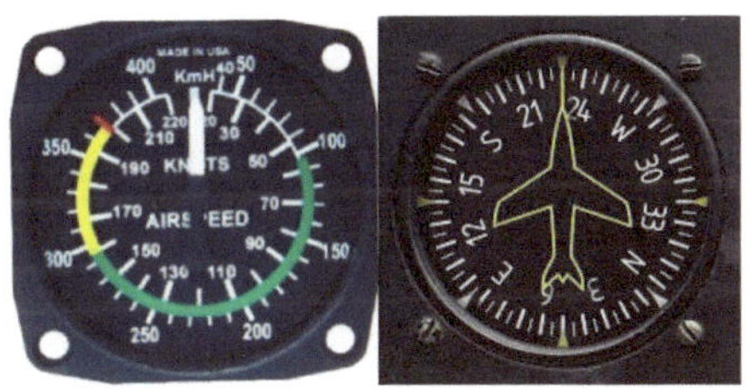

Example-2.1

An airplane is flying 15° north of east at a speed of 550mph. The wind is blowing 60° south of east at 60mph. Determine the ground speed of the airplane.

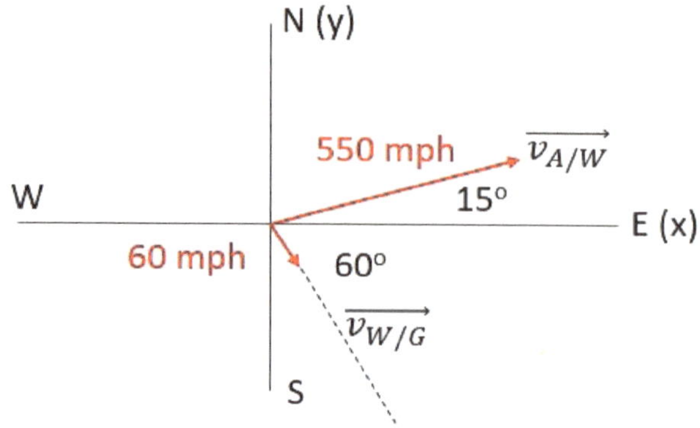

$$\overrightarrow{v_{A/W}} = 550cos15^o\hat{\imath} + 550cos75^o\hat{\jmath} = 531\hat{\imath} + 142\hat{\jmath}\ mph$$

$$\overrightarrow{v_{W/G}} = 60cos60^o\hat{\imath} + 60cos150^o\hat{\jmath} = 30\hat{\imath} - 52\hat{\jmath}\ mph$$

$$\overrightarrow{v_{A/G}} = \overrightarrow{v_{A/W}} + \overrightarrow{v_{W/G}} = 561\hat{\imath} + 90\hat{\jmath}\ mph$$

$$v_{A/G} = 568\ mph$$

Example-2.2

A pilot wants to fly an airplane 30° north of west at 450mph. The wind is blowing 20° north of east at 60mph. Determine the airspeed and direction.

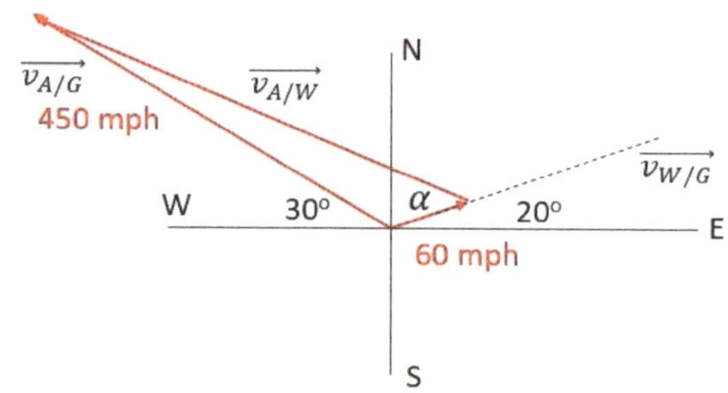

By using the cosine rule

$$v_{A/W} = \sqrt{450^2 + 60^2 - 2 \times 450 \times 60 \times cos130^o} = 491\ mph$$

By using the sine rule

$$\frac{sin\alpha}{450} = \frac{sin130^o}{491}\ ;\ \alpha = 44.6^o$$

The direction of airspeed is 24.6° north of west.

2.5 Streamlines

Streamlines are imaginary curves in a flow that are tangential to the local velocity vector at every point in the flow.

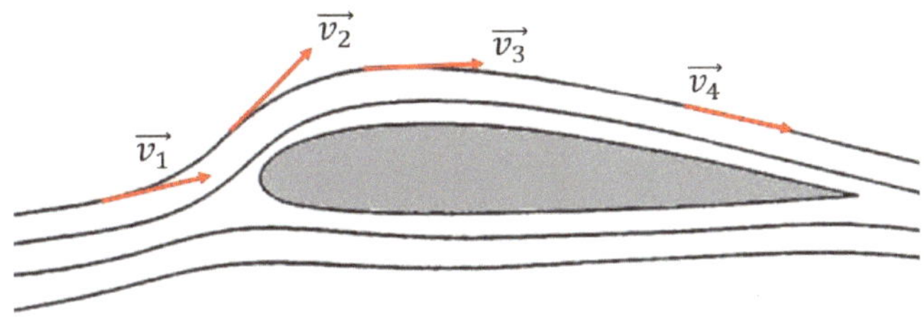

Figure 2.4: Streamlines

2.6 Conservation Laws

The three conservation laws for mass, momentum, and energy are the equations of motion in aerodynamics. Here, we will derive only the mass and the momentum conservation equations.

2.6.1 System and Control Volume

A system is a fixed volume of air. The shape, volume, velocity, and the location of a system change with time as the system flows. The mass of the system remains constant. A control volume (CV) is a region inside the flow. Systems flow across the surface of the control volume (CS), enter, and exit the CV. Different systems occupy the CV at different times.

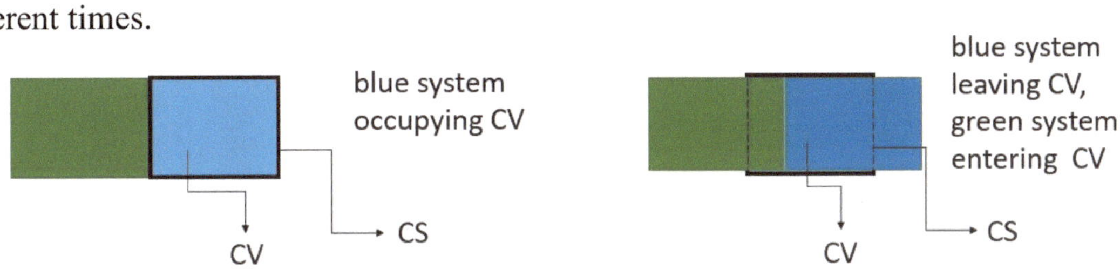

Figure 2.5: System and control volume (CV).

2.6.2 Mass Flow Rate

To derive the conservation laws, we take a rectangular parallelepiped as the CV. The centerline of the parallelepiped is a streamline that is along the x-axis. Fluid enters and exits the CV through the two surfaces that are perpendicular to the streamline. Velocity across the four surfaces, that are parallel to the streamline i.e. the x-axis, is zero.

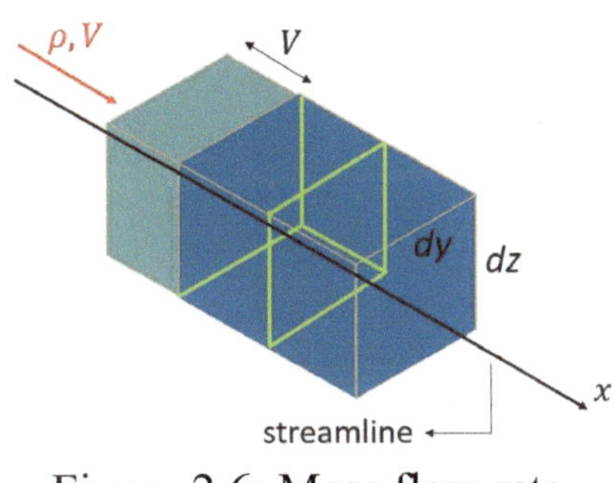

Figure 2.6: Mass flow rate

The green system is about to enter the blue CV as its right edge is at the entrance of the CV. In one second, this system of width V will travel a distance V and the system will flow into the CV. The mass of air that flows into a CV in one second is the mass inflow rate. Mass inflow rate is the mass of the green system $\rho V dydz$. Similarly, mass outflow rate is density times outflow velocity times exit area.

2.6.3 Mass Conservation

The density and velocity are ρ, v and $\rho + d\rho, v + dv$ at the inlet and outlet, respectively.

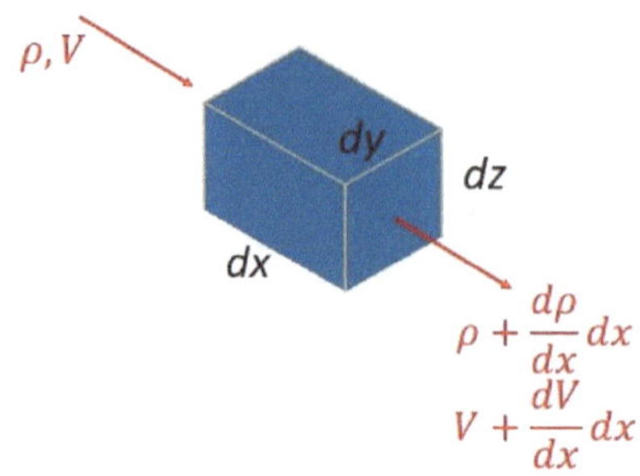

Figure 2.7: Mass conservation.

Mass inflow rate is

$$\rho V dydz$$

Mass outflow rate is

$$\left(\rho + \frac{d\rho}{dx}dx\right)\left(V + \frac{dV}{dx}dx\right)dydz = \rho V dydz + \left(V\frac{d\rho}{dx} + \rho\frac{dV}{dx}\right)dxdydz$$

For steady flow

$$Mass\ outflow\ rate\ =\ Mass\ inflow\ rate$$

$$V\frac{d\rho}{dx} + \rho\frac{dV}{dx} = 0 \tag{2.2}$$

2.6.4 Momentum Conservation

Momentum flowrate is the local mass flowrate times the local velocity.

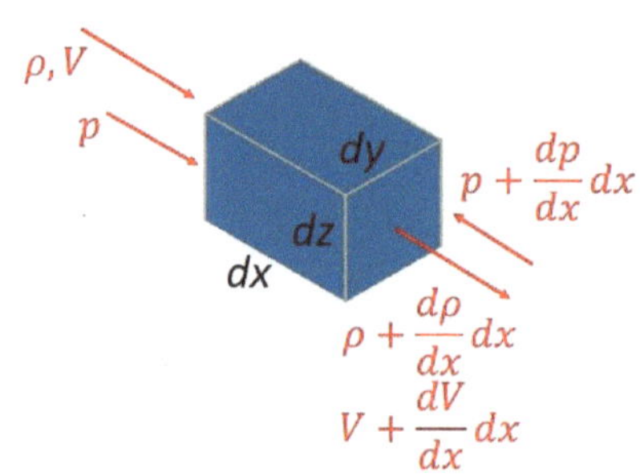

Figure 2.8: Momentum conservation

Momentum inflow rate is

$$\rho V^2 dydz$$

Momentum outflow rate is

$$\left(\rho + \frac{d\rho}{dx}dx\right)\left(V + \frac{dV}{dx}dx\right)^2 dydz = \rho V^2 dydz + \left(V^2\frac{d\rho}{dx} + 2\rho V\frac{dV}{dx}\right)dxdydz$$

By using Eqn. (2.2), momentum outflow rate is

$$\rho V^2 dydz + \rho V\frac{dV}{dx}dxdydz$$

Pressure is a squeezing force per unit area *(N/m², lb/ft²)* that acts into a surface and is perpendicular to the surface. The *x* component of force on the air in the CV is

$$pdydz - \left(p + \frac{dp}{dx}dx\right)dydz = -\frac{dp}{dx}dxdydz$$

For steady flow

$$Momentum\ inflow\ rate + x\ component\ of\ force = Momentum\ outflow\ rate$$

We neglect the force from gravity in the momentum equation because specific weight of air is small and the airspeed is large. In addition, we neglect the effect of viscosity.

$$\rho V^2 dydz - \frac{dp}{dx}dxdydz = \rho V^2 dydz + \rho V\frac{dV}{dx}dxdydz$$

Or

$$V\frac{dV}{dx} + \frac{1}{\rho}\frac{dp}{dx} = 0 \qquad (2.3)$$

When density does not change with pressure, we can integrate the momentum conservation equation along the streamline i.e. the *x*-axis to obtain

$$p + \frac{1}{2}\rho V^2 = constant \qquad (2.4)$$

This is the Bernoulli Equation. Pressure p is the static pressure in the flow that is measurable with a barometer. The second term in the Bernoulli Equation is the dynamic pressure Q . The unit of Q is *N/m² (lb/ft²)*.

$$Q = \frac{1}{2}\rho V^2 \qquad (2.5)$$

Observations

- Static and dynamic pressures add up to a constant.
- For high velocity, dynamic pressure is high and static pressure is low.
- For low velocity, dynamic pressure is low and static pressure is high.

Caution

- *We derived the Bernoulli Equation by using a control volume that has its axis along a streamline. Therefore, the Bernoulli Equations is valid only along a streamline.*
- *We assumed that density does not change with pressure or the fluid is incompressible. This assumption is valid for low speed flow of air. In the rest of this Chapter, we will assume that air is incompressible.*

The Bernoulli Equation between two points on a streamline is

$$p_1 + \frac{1}{2}\rho V_1^2 = p_2 + \frac{1}{2}\rho V_2^2 \tag{2.6}$$

Example-2.3

In an incompressible flow past an airfoil, the pressure, density, and velocity far ahead of the airfoil are 973.3 lb/ft², 0.00127 slug/ft³, and 300 mi/h. Determine the pressure at a point on the airfoil where the velocity is 340 mi/h.

$$V_1 = \frac{300 \times 5280}{3600} = 440\frac{ft}{s} \quad ; \quad V_2 = \frac{340 \times 5280}{3600} = 498.7\frac{ft}{s}$$

From Bernoulli Equation

$$973.3 + 0.5 \times 0.00127 \times 440^2 = p_2 + 0.5 \times 0.00127 \times 498.7^2$$

$$p_2 = 938.3\frac{lb}{ft^2}$$

2.7 Pressure Coefficient

Consider an airfoil with airspeed V . Far ahead of the airfoil, the pressure is atmospheric p_{Atm} . Points U and L are on the upper and lower surfaces of the airfoil.

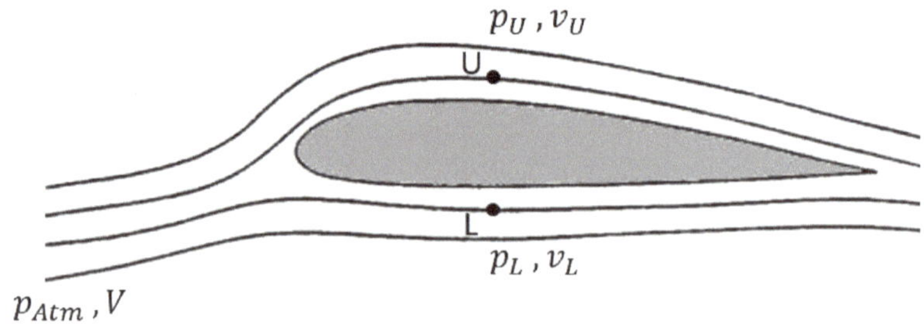

Figure 2.9: Pressure coefficient

The Bernoulli Equation (2.6) between upstream and point U is

$$p_{Atm} + \frac{1}{2}\rho V^2 = p_U + \frac{1}{2}\rho v_U^2$$

$$p_U - p_{Atm} = \frac{1}{2}\rho V^2 - \frac{1}{2}\rho v_U^2 = \frac{1}{2}\rho V^2\left(1 - \frac{v_U^2}{V^2}\right)$$

$$c_{pU} = \frac{p_U - p_{Atm}}{Q} = \left(1 - \frac{v_U^2}{V^2}\right) \tag{2.7}$$

The quantity c_{pU} is the pressure coefficient at the upper surface. For upper surface,

$$v_U > V \ ; \ c_{pU} < 0 \tag{2.8}$$

Therefore, the pressure at the upper surface is less than atmospheric. Similarly, for the lower surface

$$c_{pL} = \frac{p_L - p_{Atm}}{Q} = \left(1 - \frac{v_L^2}{V^2}\right) \tag{2.9}$$

$$v_L < V \ ; \ c_{pL} > 0 \tag{2.10}$$

Therefore, the pressure at the lower surface is more than atmospheric. In the figure below, we show a typical variation of pressure coefficient on the upper and lower surfaces of an airfoil.

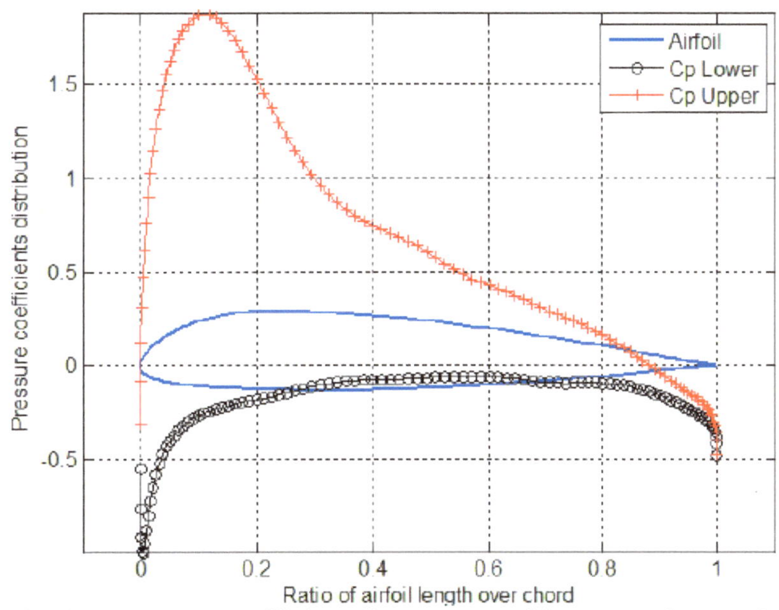

Figure 2.10: Pressure coefficient for upper and lower surfaces of airfoil.

We prefer to use an inverted ordinate where we plot a negative pressure coefficient as a positive number. We can estimate the lift force by calculating the area bounded by the upper and lower pressure coefficient plots. We will answer the question "why is $v_U > v_L$?" in Section 2.15.

2.8 Airspeed

We use the Bernoulli Equation to determine the airspeed by using a Pitot tube. We mount the Pitot tube on the fuselage of an aircraft.

Figure 2.11: Pitot tube.

Velocity inside the Pitot tube is zero because there is no outlet from the tube. By writing the Bernoulli Equation between two points, one in the approaching flow and the other inside the Pitot tube

$$p_{Atm} + \frac{1}{2}\rho V^2 = p_{Pitot} \tag{2.11}$$

We measure the pressure inside the Pitot tube and the atmospheric pressure. When we insert the density at the flight altitude, in the Bernoulli Equation, we determine the "true airspeed."

$$V_{true} = \sqrt{\frac{2(p_{Pitot} - p_{Atm})}{\rho}} \tag{2.12}$$

When we insert the density at sea level, we determine the "indicated airspeed" that is displayed on pilot's console.

$$V_{indicated} = \sqrt{\frac{2(p_{Pitot} - p_{Atm})}{\rho_{SL}}} \tag{2.13}$$

In air navigation, *knot (nautical miles per hour)* is the unit of speed.

$$1\ knot = 1.15\ mph = 1.85 \frac{km}{h}$$

<u>Example-2.4</u>

The Pitot tube attached to an airplane flying at an altitude of 20,000ft indicates a pressure of 1010 lb/ft². Determine the true and indicated airspeed.

$$p_{atm,20000} = 973.3\ {}^{lb}/_{ft^2};\ \rho_{20000} = 0.001267\ {}^{slug}/_{ft^3};\ \rho_{SL} = 0.002377\ {}^{slug}/_{ft^3}$$

$$V_{true} = \sqrt{\frac{2(1010 - 973.3)}{0.001267}} = 240.7\frac{ft}{s}$$

$$V_{indicated} = \sqrt{\frac{2(1010 - 973.3)}{0.002377}} = 175.7\frac{ft}{s}$$

2.9 Viscosity

Now that we have the basic idea behind the Bernoulli equation and inviscid flow, we will turn to viscosity.

2.9.1 No-Slip Condition

At a solid-air interface, the relative velocity between air and the solid is zero. This is the no-slip condition of fluid mechanics and is an assumption. Experimental observations and many of its consequences confirm this assumption. Imagine a layer of air between two large, parallel flat plates.

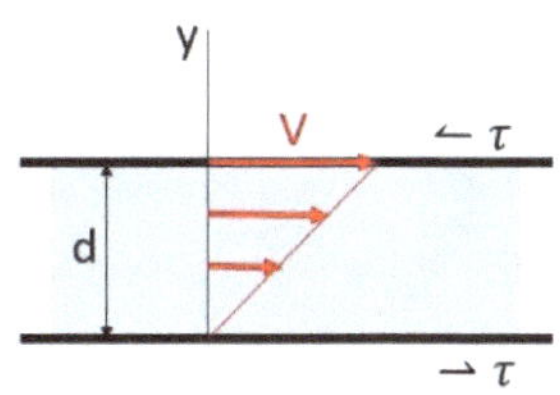

Figure 2.12: Flow of air between two parallel plates.

The top plate is moving to the right and pulling the adjacent layer of air with it due to the no-slip condition. The bottom plate is stationary and keeping the adjacent layer stationary due to the no-slip condition. The variation of velocity is linear in the gap between the plates. The slower air in the gap applies a tangential force to the left on the top plate and the faster air in the gap applies a force to the right on the bottom plate. These forces per unit area are the viscous shear stress τ *(N/m^2, lb/ft^2)*.

$$\tau = \mu\frac{V}{d}$$

The coefficient of viscosity μ is a physical property of air that depends on the temperature. Coefficient of viscosity at sea level and at an altitude of *30,000ft* is

$$\mu_{SL} = 3.737 \times 10^{-7}\frac{lb.s}{ft^2} \;\; ; \;\; \mu_{30000} = 3.107 \times 10^{-7}\frac{lb.s}{ft^2}$$

2.10 Reynolds Number

The order of magnitude of dynamics pressure is (ignoring the factor ½)

$$Q = O(\rho V^2)$$

The ratio of dynamic pressure and the viscous shear stress is the Reynolds Number.

$$Re = \frac{dynamic\ pressure}{viscous\ shear\ stress}$$

Observations

- Reynolds Number is a unit less quantity.
- It includes the fluid properties density and coefficient of viscosity, velocity of the fluid, and a characteristic length.
- When the Reynolds Number is large, dynamic pressure dominates the character of the flow.
- When the Reynolds Number is small, viscous shear stress dominates the character of the flow.

For the flow between two parallel plates, the characteristic length is the separation between the plates and the Reynolds Number is

$$Re = \frac{\rho V^2}{\frac{\mu V}{d}} = \frac{\rho V d}{\mu} \tag{2.14}$$

Observations

- For an airfoil, the characteristic length is the chord length.
- For a wing, the characteristic length is the average chord.
- For the fuselage, the characteristic length is the length of the fuselage.

Example-2.5

Determine the Reynolds Number for an airfoil with chord length of 15ft flying at speed of 250mi/h at (i) sea level and (ii) altitude of 30,000ft by ignoring the effect of altitude on the coefficient of viscosity.

$$V = \frac{250 \times 5280}{60 \times 60} = 366.7\frac{ft}{s}$$

(i) Reynolds Number at sea level

$$\rho_{SL} = 0.00238\frac{slug}{ft^3} \quad ; \quad Re = \frac{0.00238 \times 366.7 \times 15}{3.737 \times 10^{-7}} = 3.5 \times 10^7$$

(ii) Reynolds Number at 30,000ft

$$\rho_{30000} = 0.000891\frac{slug}{ft^3} \quad ; \quad Re = \frac{0.000891 \times 366.7 \times 15}{3.107 \times 10^{-7}} = 1.6 \times 10^7$$

Observations

- For a tapered wing, we define two Reynolds Numbers, one based on the root chord length and the other on the tip chord length.

- These two Reynolds Numbers play a critical role in wing design and we will discuss this in Section 2.16.

2.11 Vorticity

Due to the no-slip condition, fluid particles cannot slide on a solid surface, but roll on the solid surface, just like the balls on a pool table. This rotation of fluid particles is vorticity.

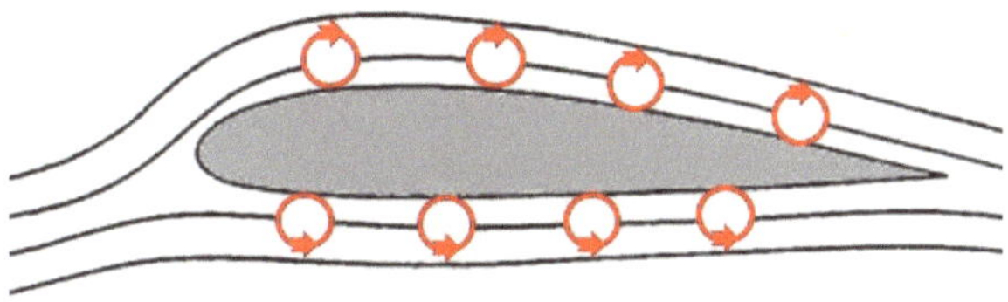

Figure 2.13: Vortices on the upper and lower surface of airfoil.

Observations

- On the top surface of the airfoil, the vorticity is clockwise and on the bottom surface, the vorticity is counter-clockwise.
- As vorticity is the effect of viscosity, it is present in a layer near a solid surface where the effect of viscosity is strong. This layer is the boundary layer.
- For large Reynolds Number, effect of viscosity is small and the boundary layer is thin.
- For small Reynolds Number, effect of viscosity is large and the boundary layer is thick.

Analysis of boundary layer yields critical information about the viscous drag on an airplane.

2.12 Boundary Layer and Skin Friction

A viscous fluid flows parallel to a flat plate with its leading edge at the origin. The velocity of the fluid far from the plate is V and velocity is zero at the plate due to the no-slip condition. The velocity of the fluid increases with y and at a sufficiently large y, the fluid velocity attains its free stream value V. Unlike the linear velocity profile in a flow between two parallel plates (Section 2.9), the velocity profile $u(y)$ is nearly parabolic.

Inside the boundary layer, the flow is rotational with vorticity. We cannot use the Bernoulli Equation in this region. Outside the boundary layer, the flow is irrotational without vorticity and the Bernoulli Equation is applicable. The boundary layer thickness is $\delta(x)$. The boundary layer thickness increases with x as more layers of fluid slow down due to viscosity, as the fluid flows downstream. We need to solve partial differential equations to obtain the relationship between the Reynolds Number and the thickness of the boundary layer. Here, we will state the results for a flat plate in viscous flow. The boundary layer thickness is (Schlichting, 1979, pg. 140)

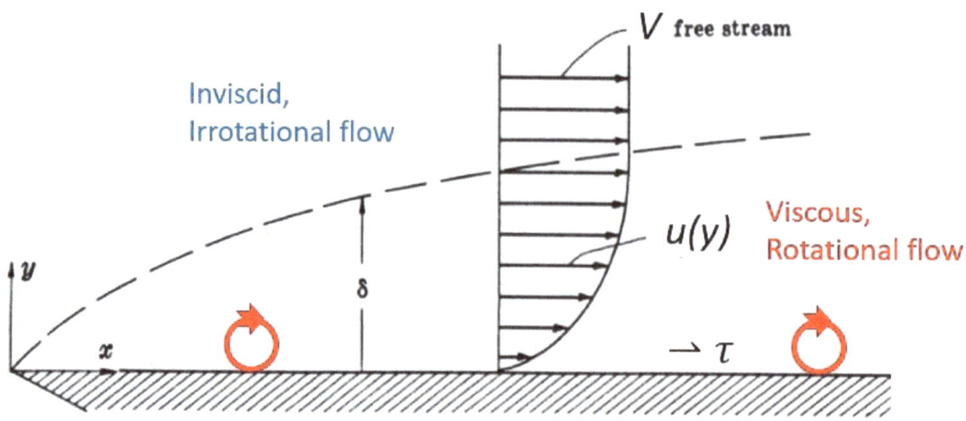

Figure 2.14: Boundary layer on a flat plate.

$$At\ y = \delta(x)\ u(y) = V$$

$$\delta(x) = \frac{5x}{\sqrt{\frac{\rho V x}{\mu}}} = \frac{5x}{\sqrt{Re_x}} \tag{2.15}$$

<u>*Example-2.6*</u>

Determine the boundary layer thickness for the two conditions of Example-2.5 at the trailing edge of an airfoil.

(i) For Reynolds number of 3.5×10^7

$$\delta(15) = \frac{5 \times 15}{\sqrt{35000000}} = 0.0127ft = 0.152in$$

(ii)For Reynolds number of 1.6×10^7

$$\delta(15) = \frac{5 \times 15}{\sqrt{16000000}} = 0.0188ft = 0.225in$$

<u>*Observation*</u>

- We can use the Bernoulli Equation in the irrotational flow outside the boundary layer, but not in the rotational flow inside the boundary layer. However, the solutions of Example-2.6 show that the boundary layer is so thin that almost the entire flow is irrotational. In this situation, our calculation of pressure coefficient (see Section 2.7) from the Bernoulli Equation has acceptable accuracy. However, we will see in Section 2.15 that ignoring the boundary layer leads to an incorrect description of the flow past an airfoil.

The viscous shear stress at the plate is (Schlichting, 1979, pg. 138)

$$\tau(x) = \mu\left(\frac{\partial u}{\partial y}\right)_{y=0} = \frac{1}{2}\rho V^2 \frac{0.664}{\sqrt{Re_x}} = Q\frac{0.664}{\sqrt{Re_x}}$$

We can determine the viscous drag on one side of a flat plate of width w and length L by integrating $\tau(x)$over the entire plate.

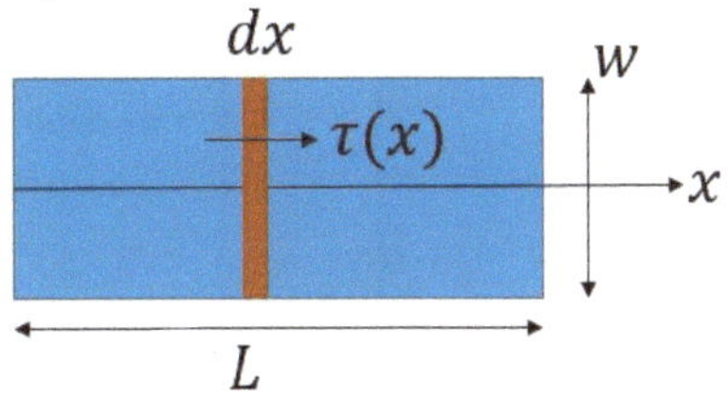

Figure 2.15: Viscous drag on a flat plate

Viscous drag D is

$$D = w\int_0^L \tau(x)dx = Qw\int_0^L \frac{0.664}{\sqrt{\frac{\rho V x}{\mu}}}dx = QwL\frac{1.328}{\sqrt{Re_L}}$$

We define a unit less friction coefficient as

$$C_f = \frac{D}{QwL} = \frac{1.328}{\sqrt{Re_L}} \quad ; \quad Re_L = \frac{\rho V L}{\mu} \tag{2.16}$$

2.13 Friction Coefficient in Turbulent Flow

Wind tunnel experiments show that for $Re < 10{,}000$, the flow is laminar in parallel layers with no lateral mixing, no crosscurrents perpendicular to the direction of the flow, and no eddies or swirls. For $Re > 10{,}000$, the flow is turbulent with lateral mixing, cross currents, and eddies in the flow.

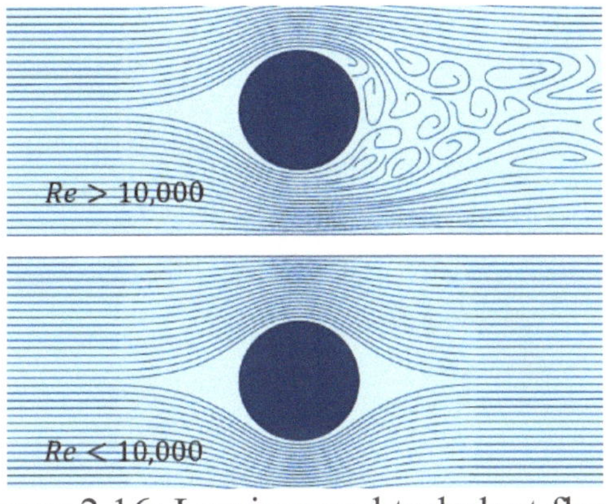

Figure 2.16: Laminar and turbulent flows.

Within the boundary layer, the lateral mixing affects the thickness of the boundary layer and the speed of air in various layers.

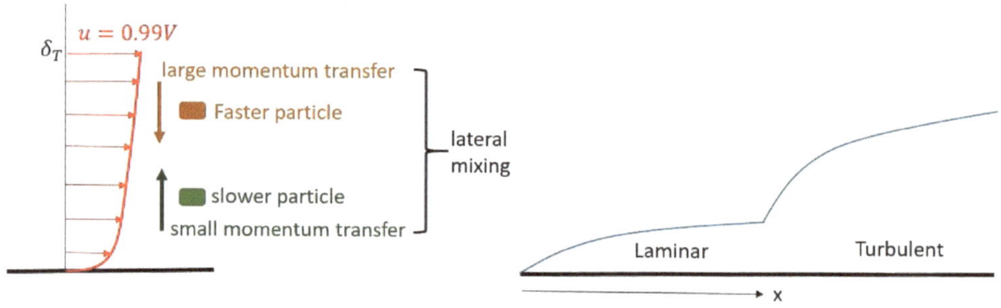

Figure 2.17: Laminar and turbulent boundary layers.

In lateral mixing, faster particles from the outer layers add large momentum to the inner layers and slower particles from the inner layers add small momentum to the outer layers. The outer layers have large momentum outflow and small momentum inflow. The inner layers have small momentum outflow and large momentum inflow. This transfer of momentum accelerates the inner layers and decelerates the outer layers.

Observation

- Deceleration of the outer layers makes turbulent boundary layer thicker.
- Acceleration of the inner layers increases the momentum and kinetic energy content in these layers.

The flow on an airfoil is turbulent as the Reynolds Number is of the order of millions. For turbulent flows, the friction coefficient of Eqn. (2.16) is inaccurate because we excluded the lateral mixing in turbulent flows. In our calculations, we will use the friction coefficient for turbulent flow on a smooth flat plate with Reynolds Number based on chord *c* is (Schlichting, 1979, pg. 641). We will correct this flat plate formula for the thickness of the airfoil in Section 2.20.8.

$$C_f = \frac{0.455}{\{log_{10}Re\}^{2.58}} \;\; ; \; Re = \frac{\rho V c}{\mu} \tag{2.17}$$

Example-2.7

Determine the friction coefficient for the two conditions of Example-2.5.

(i) For Reynolds Number of 3.5×10^7

$$C_f = \frac{0.455}{\{log_{10}(3.5 \times 10^7)\}^{2.58}} = 0.002476$$

(ii)For Reynolds Number of 1.6×10^7

$$C_f = \frac{0.455}{\{log_{10}(1.6 \times 10^7)\}^{2.58}} = 0.002789$$

Observation

- Reynolds Number decreases with the increase in altitude.
- Boundary layer thickness increases with the increase in altitude.
- Friction coefficient increases with the increase in altitude.

The figure below shows the variation of friction coefficient with Reynolds Number.

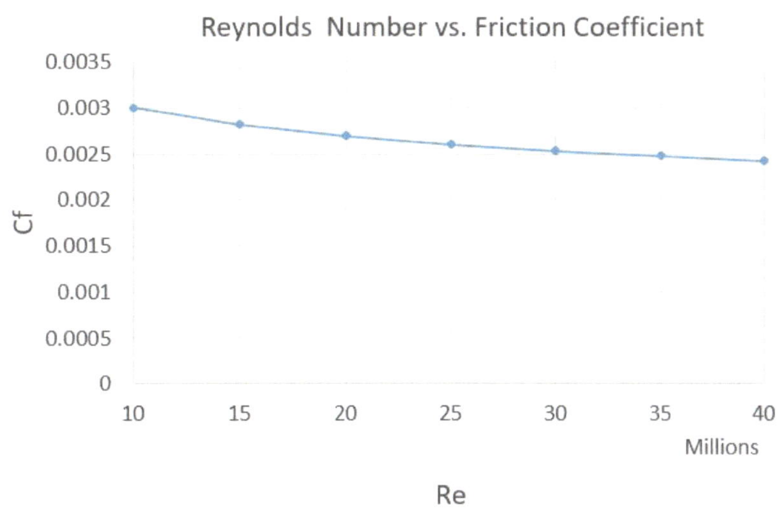

Figure 2.18: Variation of friction coefficient with Reynolds Number.

Observation

- Friction coefficient is a slowly varying function of Reynolds Number.
- For an airplane wing in cruise, we will use the value *0.00275* for friction coefficient for preliminary calculations.

2.14 Point Vortex

In Section 2.11, we saw how the no-slip condition produces vorticity as rotating particles at the air-solid interface. A rotating particle induces a flow around it with circular streamlines. The mathematical theory of these circular streamlines is the "point vortex model."

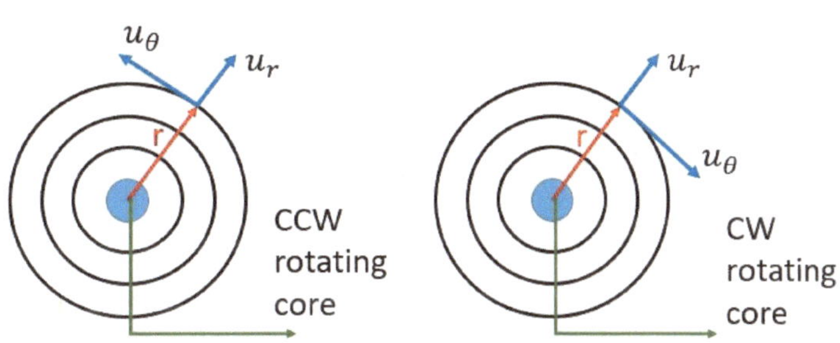

Figure 2.19: Point vortices.

In this model, for a point vortex of strength γ , the velocity components are

$$u_r = 0 \; ; \; u_\theta = \frac{\gamma}{2\pi r}$$

The effect of the rotating core on the surrounding fluid diminishes as *r* increases i.e. when we move away from the core. We define the strength of a point vortex by the amount of circulation the vortex produces where the circulation of the vortex core is the velocity integrated along a circular streamline

$$Strength\ of\ Vortex = Circulation = \int_0^{2\pi} u_\theta r d\theta = \gamma$$

The unit of circulation is ft^2/s. The net vortex strength of all the vortices

$$\Gamma = \sum_{all\ vortices} \gamma$$

The net vortex strength is unique and depends on the shape of the airfoil and the angle of attack that is the angle between the chord and the free stream velocity *V*. The unique Γ is determined by using the Kutta Condition.

2.15 Kutta Condition

To solve the problem of flow of air past an airfoil, we need to write a computer program that is beyond the scope of this book. Such a computer program for the *inviscid flow* past an airfoil *without including vorticity* yields an incorrect solution.

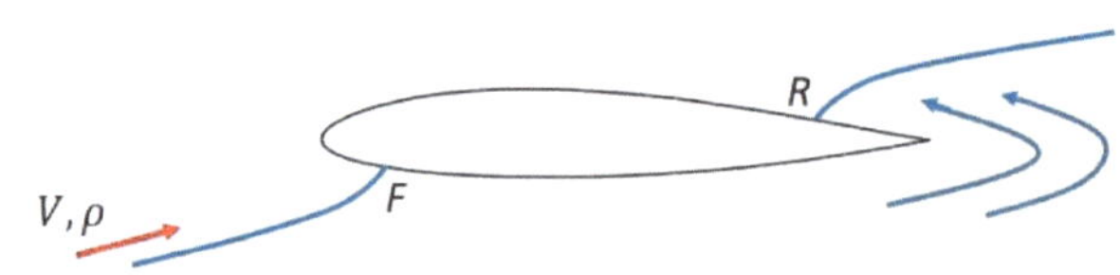

Figure 2.20: Flow past an airfoil without vorticity.

We have shown only the main incoming and outgoing streamlines and the front and rear stagnation points where the velocity is zero. In this flow, air turns around the sharp trailing edge and we find unrealistic, nearly infinite velocity near the trailing edge. Such a flow does not exist in nature. We get this incorrect flow because air is viscous and ignoring viscosity and vorticity, in the boundary layer, is an erroneous approximation. Wind tunnel tests show see the correct flow

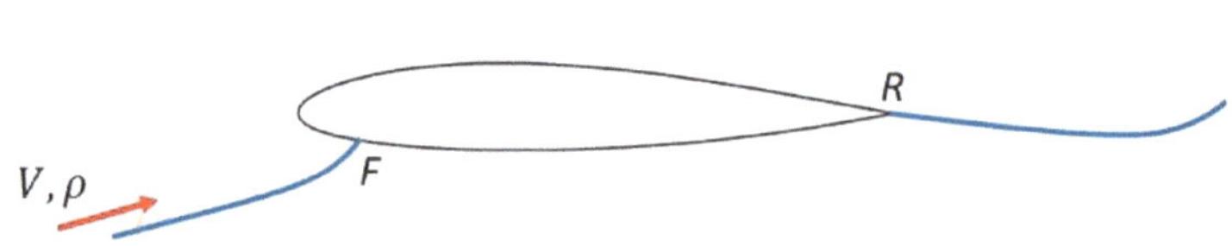

Figure 2.21: Flow past an airfoil with vorticity.

We can simulate the correct solution by distributing clockwise vorticity on the top surface and counter-clockwise vorticity on the bottom surface of the airfoil (see Figure 2.13). The correct amount of vorticity is determined by using the Kutta Condition.

In flow past airfoils, that have a sharp trailing edge, a unique amount of vorticity is added to the inviscid flow that shifts the rear stagnation point to the trailing edge.

The net vorticity in the flow is

$$\circlearrowright\Gamma = \sum_{top\ surface} \circlearrowright\Gamma - \sum_{bottom\ surface} \circlearrowleft\Gamma$$

When net vorticity is clockwise, we add the velocity from the vortex core to the flow velocity on the top surface and subtract the velocity from the vortex core from the flow velocity on the bottom surface. Thus

$$v_U > v_L$$

The right amount of clockwise vorticity makes v_U faster by the right amount such that it pushes the rear stagnation point from the top surface to the trailing edge. The Kutta Condition makes the amount of vorticity unique for an airfoil for a given angle of attack.

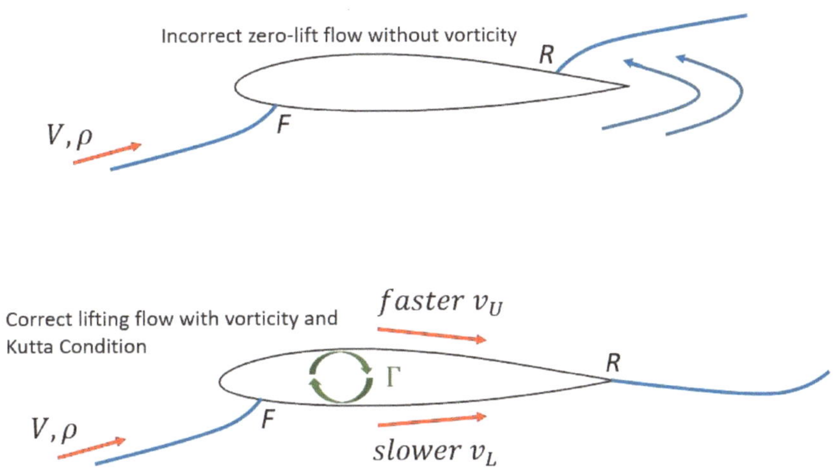

Figure 2.22: Kutta Condition

<u>*Observation*</u>

- Lift of an airfoil is $l = \rho V\Gamma$. The lift is the force on a wing of unit span and has the unit *N/m* or *lb/ft*.
- Larger camber produces larger $\circlearrowright\Gamma$ and larger lift.

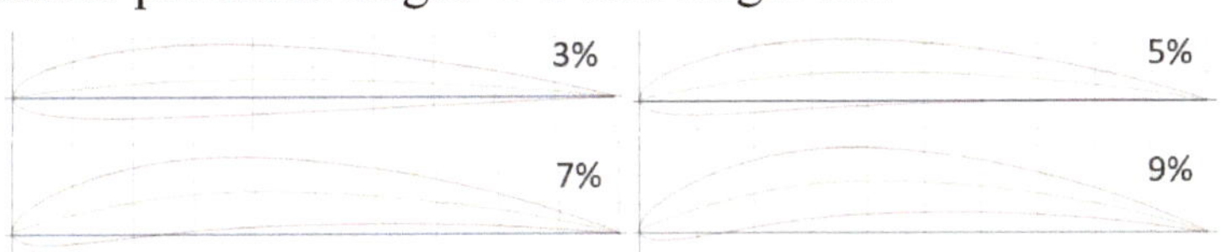

Figure 2.23: Cambered airfoils.

- Deflected flap has the same effect as increased camber.

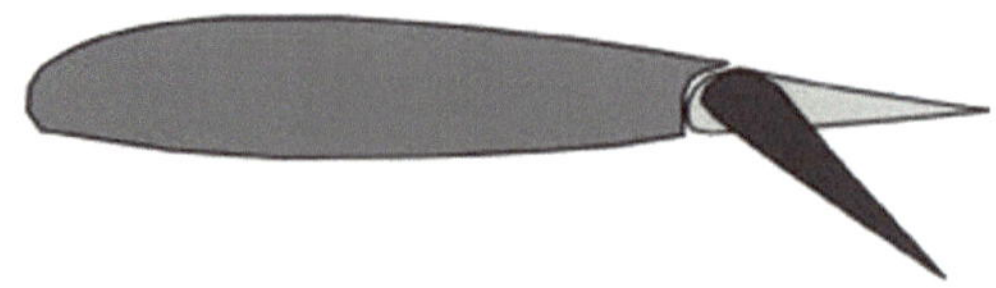

Figure 2.24: Deflected flap.

- Larger angle of attack produces larger ↻ Γ and larger lift.

2.15.1 Symmetric Airfoil

For a symmetric airfoil at zero angle of attack, the inviscid flow calculation without vorticity puts the rear stagnation point at the trailing edge. Kutta Condition does not require any added vorticity in the flow. Hence, net vorticity and lift are zero.

Figure 2.25: Zero-lift symmetric airfoil.

For a symmetric airfoil at non-zero, positive angle of attack, the inviscid flow calculation without vorticity puts the rear stagnation point on the top surface of the airfoil. Kutta Condition requires clockwise vorticity that pushes the rear stagnation point through a distance *d* to the trailing edge. A larger angle of attack gives a larger *d* and larger *d* requires larger clockwise vorticity. Larger clockwise vorticity produces larger lift for larger angle of attack.

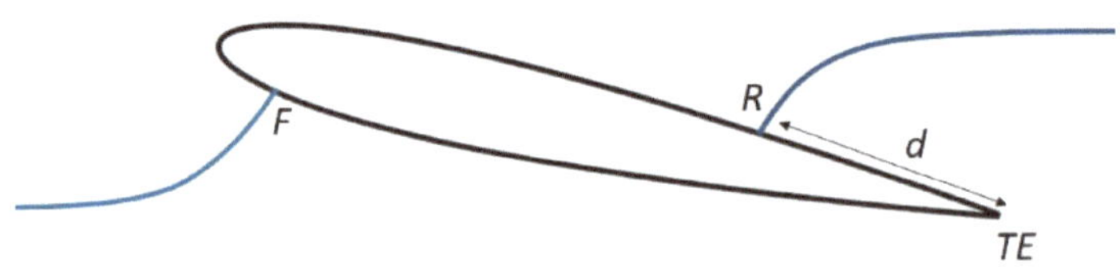

Figure 2.26: Symmetric airfoil at an angle of attack without vorticity.

2.15.2 Cambered Airfoil

For a cambered airfoil at zero angle of attack, the inviscid flow calculation without vorticity puts the rear stagnation point on the top surface of the airfoil. Kutta Condition requires clockwise vorticity that pushes the rear stagnation point through a distance *d* to the trailing edge. Therefore, a cambered airfoil produces lift even at zero angle of attack.

Figure 2.27: Cambered airfoil at zero angle of attack without vorticity.

For a cambered airfoil at non-zero, positive angle of attack, the inviscid flow calculation without vorticity puts the rear stagnation point on the top surface of the airfoil. Kutta Condition requires clockwise vorticity that pushes the rear stagnation point through a distance *d* to the trailing edge. A larger angle of attack gives a larger *d* and larger *d* requires larger clockwise vorticity. Larger clockwise vorticity produces larger lift for larger angle of attack.

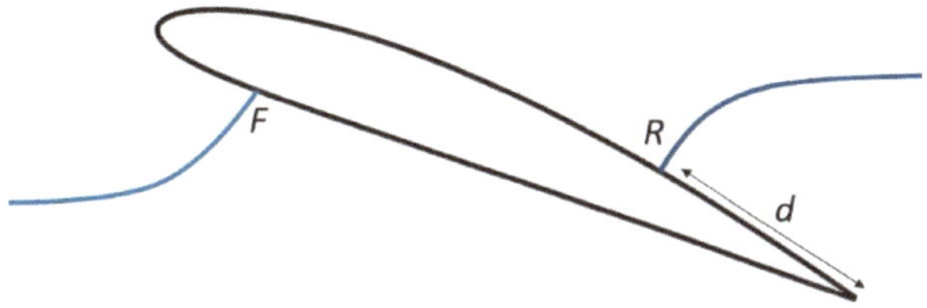

Figure 2.28: Cambered airfoil at non-zero angle of attack without vorticity.

For a cambered airfoil at zero-lift angle of attack, the inviscid flow calculation without vorticity puts the rear stagnation point at the trailing edge. Kutta Condition does not require any added vorticity in the flow. Hence, net vorticity and lift are zero.

Figure 2.29: Cambered airfoil at zero lift angle of attack.

2.16 Separation and Stall

For an airfoil at an angle of attack, velocity at *P* on the top surface is high and pressure is low, from Bernoulli Equation. Velocity at the rear stagnation point *R* is zero and pressure is high, from Bernoulli Equation. When air travels from *P* to *R,* it has to climb a pressure hill and overcome viscous shear stress.

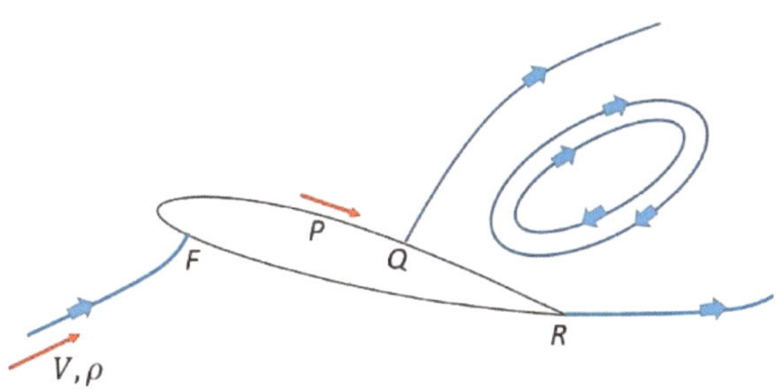

Figure 2.30: Separation

When the angle of attack increases, velocity at P increases, pressure decreases, and the pressure hill from P to R becomes steeper. At "stall" angle of attack, air cannot negotiate the shear stress and climb the pressure hill. The main flow separates from the top surface of the airfoil at point Q. The high pressure at R drives a reversed flow toward the lower pressure at Q. The reversed flow region contains a slow spinning eddy. The pressure in the region QR rises and the airfoil loses lift. Velocity profiles near and beyond the point of separation show the reversed flow.

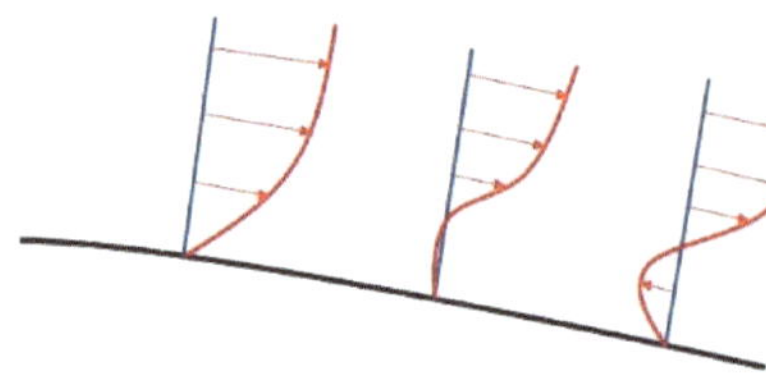

Figure 2.31: Reversed flow in separation.

In turbulent flow, the inner layers have enhanced momentum and kinetic energy (see Section 2.13). These layers can climb a hill of higher pressure compared to laminar flow.

Observation

- Turbulent flows or flows with higher Reynolds Number remain attached and we observe a delayed separation.
- In a tapered wing, the root chord is larger than the tip chord. Consequently, the root Reynolds Number is larger than the tip Reynolds Number. Thus, we observe delayed separation at the root section compared to the separation at the tip.

2.17 Lift, Moment, and Drag

To determine the lift, drag, and moment on an airfoil, we will take the origin at the leading edge of the airfoil, align the *x-axis* along the chord, the *y-axis* perpendicular to the chord, and take an elemental area dA on the surface of the airfoil.

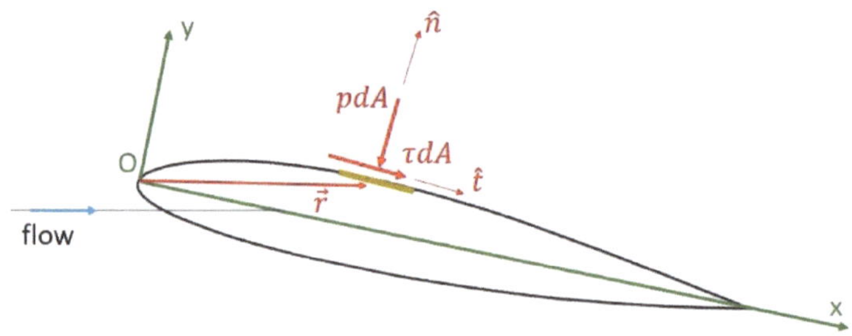

Figure 2.32: Pressure and shear stress on airfoil.

The force on the elemental area dA on the airfoil is

$$\overrightarrow{dF} = -pdA\,\hat{n} + \tau dA\,\hat{t}$$

The unit normal and unit tangent vectors on area dA are $\hat{n}$ and $\hat{t}$. Pressure p acts perpendicular to the area element and viscous shear stress τ acts tangential to the area element. We can also draw a $\vec{r}$ from the origin to the area dA to calculate

$$\overrightarrow{dm_O} = \vec{r} \times \overrightarrow{dF}$$

By integrating over the entire surface of the airfoil, we find drag, lift, and m_O.

$$\vec{F} = \int \overrightarrow{dF} = F_x\hat{x} + F_y\hat{y} \ ; \ \ \overrightarrow{m_O} = \int \overrightarrow{dm_O} = m_O\hat{z}$$

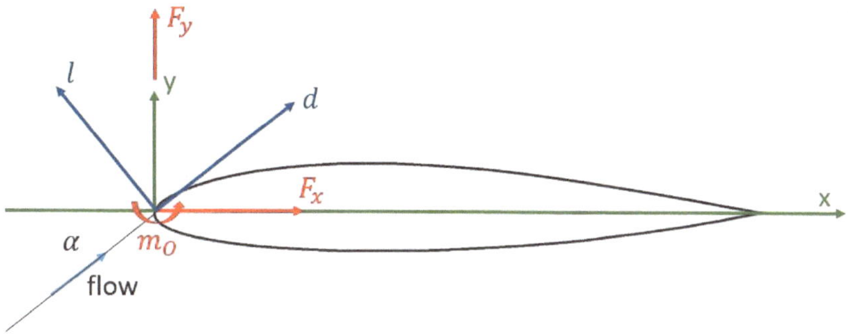

Figure 2.33: Forces and moment on an airfoil.

Observation

- l is the lift of the airfoil and is perpendicular to the flow.
- d is the drag on the airfoil and is along the flow.
- m_o is the pitching at the leading edge of the airfoil

For small angle of attack α

$$F_x = dcos\alpha - lsin\alpha \approx d \ ; \ F_y = lcos\alpha + dsin\alpha \approx l$$

2.17.1 Pitching Moment at Quarter Chord

We can measure or calculate force and moment on an airfoil in two equivalent force-moment system.

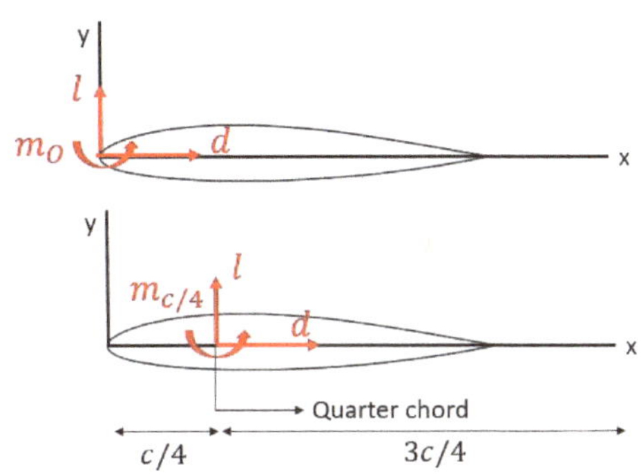

Figure 2.34: Lift, drag, and moment at quarter chord.

We can write a relationship among the moments in the two systems.

$$m_O = m_{c/4} + \frac{1}{4}cl \tag{2.18}$$

The moment at the quarter chord is important because we measure this moment in wind tunnel testing by mounting a JR3-device (a multi-axis load cell) at the quarter chord point. This device measures all the forces and moments on a wing.

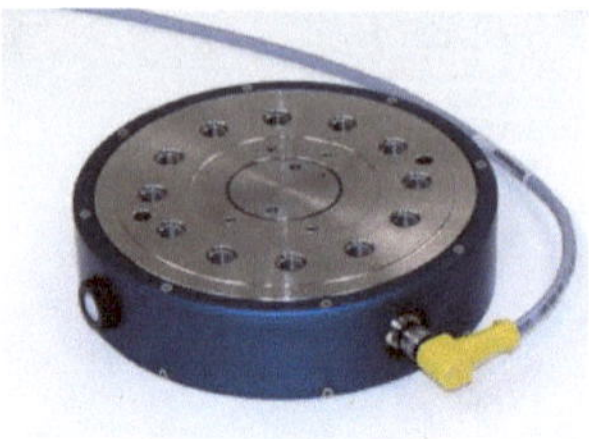

Figure 2.35: Multi-axis load cell.

An airfoil is a slice of unit length of a wing. Therefore, the units of lift and drag are *N/m (lb/ft)* and the unit of pitching moment is *N.m/m (lb.ft/ft)*. We define three unit-less quantities as

$$C_l = \frac{l}{Qc} \;;\; C_d = \frac{d}{Qc} \;;\; C_m = \frac{m_{c/4}}{Qc^2} \tag{2.19}$$

These quantities are known as lift, drag, and moment coefficients of an airfoil (note the lower-case subscripts), c is the chord, and Q is the dynamic pressure.

2.17.2 Center of Pressure

We can measure or calculate force and moment on an airfoil in two equivalent force-moment system.

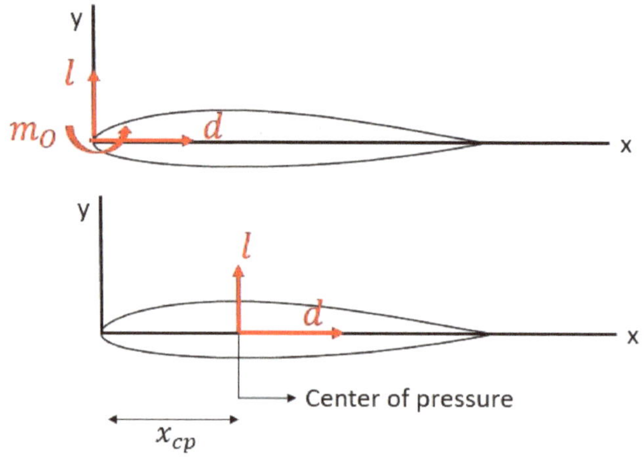

Figure 2.36: Lift and drag at center of pressure.

The center of pressure is a point on the chord where the moment on the airfoil is zero.

$$m_O = x_{cp} l \tag{2.20}$$

By combining Eqns. (2.18), (2.19), and (2.20)

$$x_{cp} = \frac{m_o}{l} = \frac{1}{4}c + \frac{m_{c/4}}{l} = \frac{1}{4}c + \frac{C_m Q c^2}{C_l Q c} = c\left(\frac{1}{4} + \frac{C_m}{C_l}\right) \tag{2.21}$$

2.17.3 Aerodynamic Center

The moment at a general location x on the chord is calculated from the equation

$$m_O = m(x) + lx \ ; \quad m_{c/4} + \frac{1}{4}cl = m(x) + lx$$

Or

$$m(x) = m_{c/4} + l\left(\frac{1}{4}c - x\right) \tag{2.22}$$

The x location where moment is independent of the angle of attack is the aerodynamic center.

$$\frac{\partial m(x)}{\partial \alpha} = \frac{\partial m_{c/4}}{\partial \alpha} + \frac{\partial l}{\partial \alpha}\left(\frac{1}{4}c - x_{ac}\right) = 0$$

We write this equation in terms of the lift and moment coefficients.

$$Qc^2 \frac{\partial C_m}{\partial \alpha} + Qc \frac{\partial C_l}{\partial \alpha}\left(\frac{1}{4}c - x_{ac}\right) = 0$$

By simplifying

$$x_{ac} = \frac{1}{4}c + c\frac{\frac{\partial C_m}{\partial \alpha}}{\frac{\partial C_l}{\partial \alpha}} \ ; \quad m_{ac} = m_{c/4} + l\left(\frac{1}{4}c - x_{ac}\right) \tag{2.23}$$

$$C_{m,ac} = \frac{m_{ac}}{Qc^2} \tag{2.24}$$

<u>*Example-2.8*</u>

An airfoil has its aerodynamic center located at 25.5% chord; the slope of the lift coefficient curve is 0.11 deg^{-1}, and a zero lift angle of attack of -2.75 degrees. The moment coefficient at the quarter chord is -0.06 when the angle of attack is 7.5^o. Find the moment coefficient at the quarter chord at 0° angle of attack and the location of the center of pressure at 0° angle of attack.

From Eqn. (2.23),

$$\frac{x_{ac}}{c} = 0.255 = 0.25 + \frac{\frac{\partial C_m}{\partial \alpha}}{0.11} \; ; \; \frac{\partial C_m}{\partial \alpha} = 5.5 \times 10^{-4}\, deg^{-1}$$

$$\frac{C_{m,7.5^o} - C_{m,0^o}}{7.5} = 5.5 \times 10^{-4}\, deg^{-1} \; ; \; \frac{-0.06 - C_{m,0^o}}{7.5} = 5.5 \times 10^{-4}\, deg^{-1}$$

Hence

$$C_{m,0^o} = -0.06413$$

$$C_{l,0^o} = 0.11\{0 - (-2.75)\} = 0.3025$$

From Eqn. (2.21),

$$\frac{x_{cp,0^o}}{c} = 0.25 + \frac{C_{m,0^o}}{C_{l,0^o}} = 0.25 + \frac{-0.06413}{0.3025} = 0.038$$

Example-2.9

For an airfoil with 12ft chord, the moment coefficient at quarter chord is -0.06. Determine the moment at sea level and at an airspeed of 200 mi/h.

$$V = \frac{200 \times 5280}{3600} = 293.3\, ft/s \; ; \; Q = 0.5 \times 0.00238 \times 293.3^2 = 102.4 \frac{lb}{ft^2}$$

$$m_{c/4} = -0.06 \times 102.4 \times 12^2 = -884.7\, lb.\frac{ft}{ft}$$

2.18 Airfoil Characteristics

2.18.1 Airfoil Geometry

We define the airfoil geometry by a camber function $y_c(x)$and a thickness function $y_t(x)$. We align the *x-axis* with the chord with the origin at the leading edge. We construct the upper and lower surfaces of the airfoil as

$$y_U(x) = y_c(x) + y_t(x) \; ; \; y_L(x) = y_c(x) - y_t(x)$$

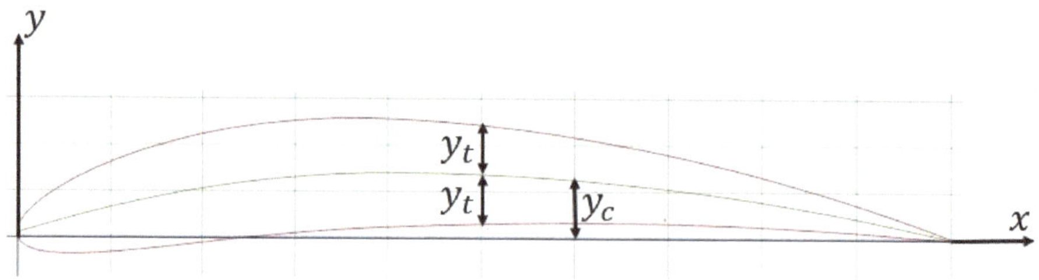

Figure 2.37: Cambered airfoil.

One popular class of airfoils is NACA 4-digit airfoils, such as NACA 2412. This airfoil has a maximum camber of *2%* located at *40%* chord from the leading edge and a maximum thickness of *12%* chord. NACA 0012 is a symmetric airfoil with no camber and *12%* maximum thickness. Thickness of the airfoil gives volume to the wing and is an important design parameter when we want to store fuel in the wing. Thicker airfoils have more drag due to viscosity.

The figures below show the variation of the lift coefficient, the drag coefficient, the moment coefficient at quarter chord, and the moment coefficient at aerodynamic center with the angle of attack for a NACA 2412 airfoil (Abbott et al., 1945).

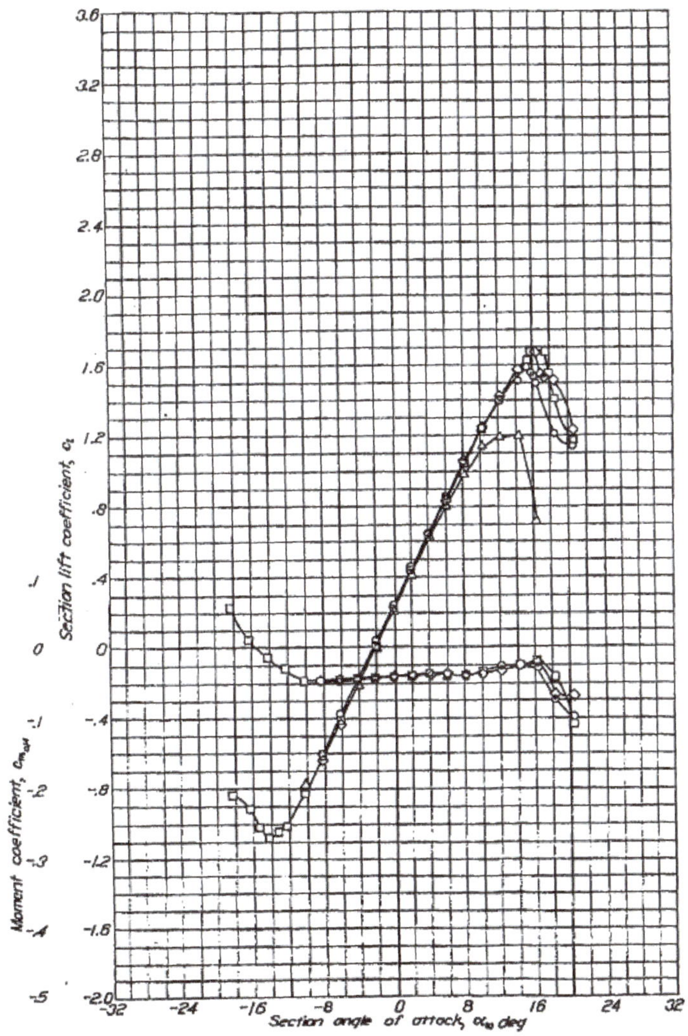

Figure 2.38: Lift and moment coefficient (quarter chord) of NACA 2412.

2.18.2 Variation of C_l

- Variation of lift coefficient with angle of attack is linear over a range of angle of attack. In this range, the slope of the lift curve for an airfoil increases as the camber of the airfoil increases. Lift curve for the NACA-2412, shown in the figure, is linear between *-10º* to *8º*. The slope of the lift curve is determined, from the graph, as

$$a_0 = \frac{1.0 - (-0.8)}{18} = 0.1\ per\ degree$$

$$a_0 = \frac{0.1 \times 180}{\pi} = 5.73 \; per \; radian$$

For all the airfoils that are in use today, the lift curve slopes are close to this value. When you do not have access to the lift curve, you can do a rough calculation by using the lift curve slope as 2π per radian (*0.11* per degree). This value of the lift curve slope is the result from the "Thin Airfoil Theory."

- The angle of attack at zero lift is $\alpha_{L=0}$. For cambered airfoils, this angle is negative and decreases with camber. For symmetric airfoils, this angle is zero. For NACA2412, the zero-lift angle of attack is *-2°*. For positive lift, the operating angle of attack is larger than the zero-lift angle of attack.
- When the angle of attack increases, lift coefficient reaches a maximum and then drops. This drop is due to stall (see Section 2.16). The stall angle of attack is larger for larger Reynolds Number (see Section 2.16).

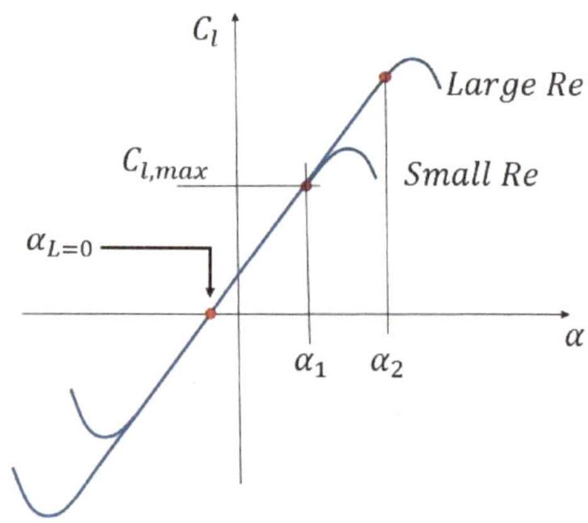

Figure 2.39: Stall angles for small and large Reynolds Number

In a wing, the tip chord is smaller than the root chord and consequently tip Reynolds Number is smaller than root Reynolds Number. The stall angle for the tip is smaller than the stall angle for the root. We give the wing a twist down to keep the tip at a smaller angle of attack. As taper increases (tip gets smaller), the twist down angle increases. Typical tip down twist is *3°* to *4°* (see Section 2.20.3).

- For an untwisted wing, the upper limit of angle of attack is α_1 to prevent tip stall. Corresponding to this angle of attack the maximum lift coefficient is $C_{l,max}$. This value is important because during landing when we want high lift at low airspeed, the aircraft will pitch up and the airfoil will fly at angle of attack α_1 . We also employ flaps and leading-edge slats to increase the amount of lift at lower airspeeds. This allows the pilot to fly at a lower angle of attack at low speed and lowering the nose increases pilot's visibility of the runway environment.
- The operating range of angle of attack is between $\alpha_{L=0}$ and α_1. In this range, we can determine the lift coefficient from the equation

$$C_l = a_0(\alpha - \alpha_{L=0}) \tag{2.25}$$

- Angle of incidence is the angle between the root chord of the wing and the fuselage reference line (FRL). For a wing with angle of incidence i_W, fuselage pitch-up angle α_{FLR}, and wing angle of attack α, the alignments of lift, drag, and moment at the aerodynamic center are:

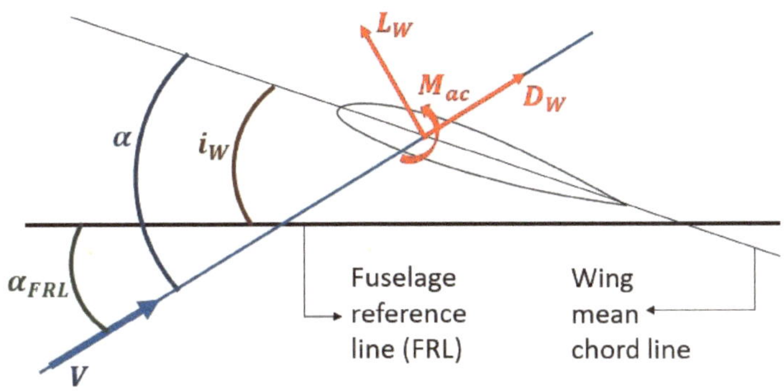

Figure 2.40: Angle of incidence and angle of attack.

- A negative camber turns an airfoil upside down. In some wing-tail arrangements, the airfoil in the tail has negative camber.

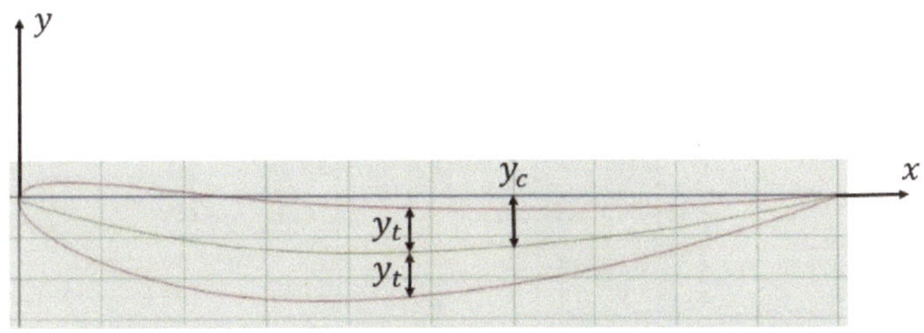

Figure 2.41: Airfoil with negative camber

The zero-lift angle of attack is negative for airfoils with positive camber ($\alpha_{L=0} < 0 \; for \; y_c > 0$). The zero-lift angle of attack is positive for airfoils with negative camber ($\alpha_{L=0} > 0 \; for \; y_c < 0$).

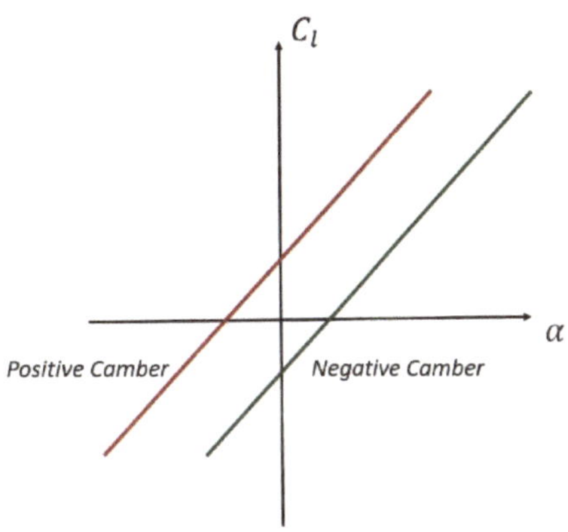

Figure 2.42: Comparison of positive and negative camber

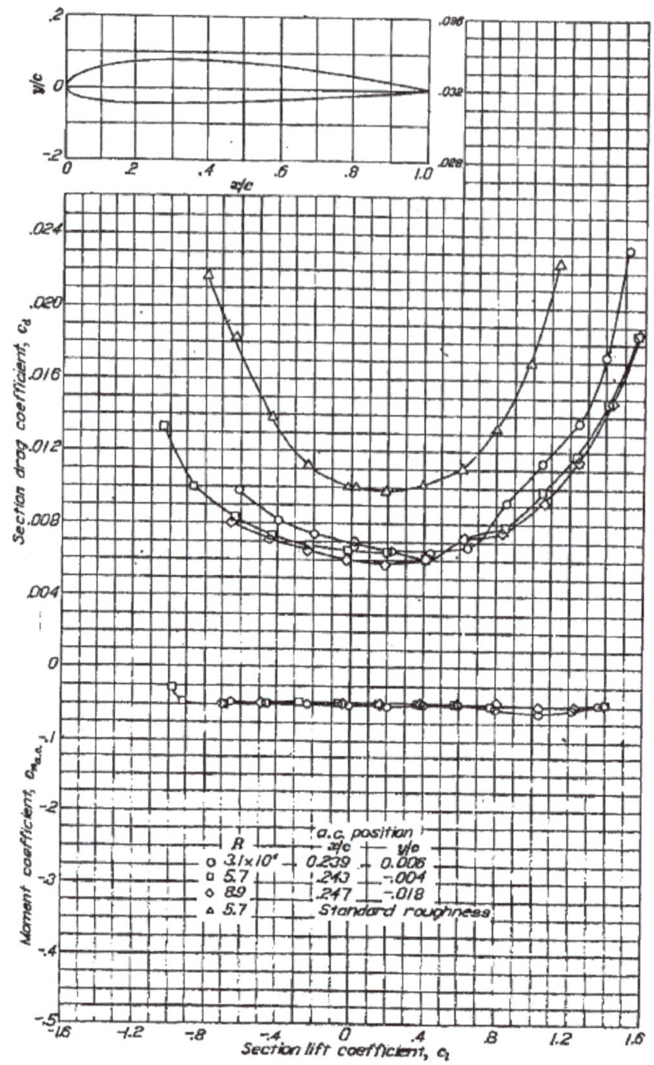

Figure 2.43: Drag and moment coefficient (aerodynamic center) of NACA 2412.

2.18.3 *Variation of* C_d

- The drag coefficient plot has the lift coefficient as the abscissa.
- The drag coefficient curve has the shape of a bucket. The bottom of this bucket corresponds to zero lift. At this bottom, the drag is entirely due to viscosity. This drag is the parasite drag (see Section 2.20.5).
- The drag coefficient has a near quadratic dependence on the lift coefficient. This quadratic part of drag is the induced drag (see Section 2.20.4).

2.18.4 *Variation of* $C_{m,ac}$

- Typically, moment at the quarter-chord point or at the aerodynamic center is clockwise and hence, the moment coefficient is negative.
- The moment coefficient plot has the lift coefficient as the abscissa.
- The moment coefficient at the aerodynamic center is nearly constant and that is consistent with the definition of the aerodynamic center.

These characteristics are for a two-dimensional airfoil. We will now modify the characteristics to find the characteristics of a three dimensional wing.

2.19 Lifting Line

For wing analysis, we replace the point vortices (see Section 2.14) for an airfoil by line vortices. We will use a simplified physical model to explain wake, downwash, and the lift distribution on the wing.

The underlying principle in this model is that the line vortices cannot end inside the flow. If it did, then at the end of the line, we will have a discontinuity of a rotating particle with a non-rotating neighbor. Such a discontinuity does not occur in nature. We can prove this principle with rigor, but that proof is beyond the scope of this book.

In the lifting line model, we replace the wing by a lifting line (Anderson, 2017, pg. 447). We attach an infinite number of line vortices to the lifting line. To understand the physical model, we consider a simplified situation with three line vortices that bend at right angles and continue downstream to infinity giving them the shape of a horseshoe. Far away from the lifting line, the line vortices become imperceptible as they diffuse in the atmosphere. This diffusion is similar to the disappearance of a mist from an aerosol.

Figure 2.44: Diffusion

The line vortices behind the wing form the wake of the wing. In reality, infinite number of line vortices forms a vortex sheet as wake. The sense of rotation of the fluid particles on the line vortices is such that the velocity at the top of the wing is larger (see Section 2.15).

Near the tips of the wing, the vortex strength is Γ_1 and the corresponding lift is L_1 . As we move inward from the tip, the accumulated vortex strength is $\Gamma_1 + \Gamma_2$ and the corresponding lift is L_2 . The lift is maximum at the central segment that has a vortex strength of $\Gamma_1 + \Gamma_2 + \Gamma_3$ and a lift of L_3 . For three line vortices, the lift distribution is three discreet rectangles. When we build a physical model with many line vortices, the lift distribution smoothens to near elliptic.

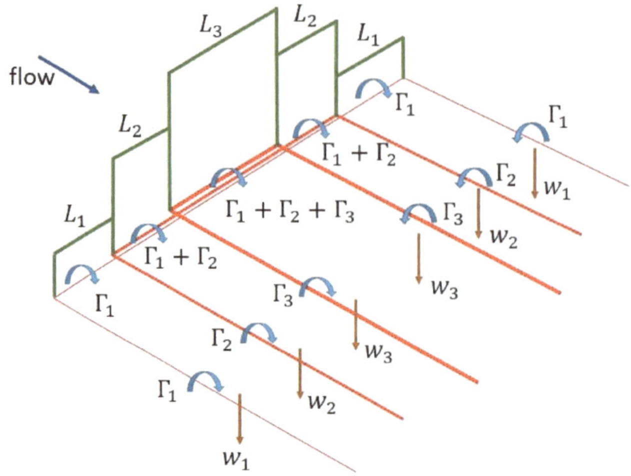

Figure 2.45: Lifting line.

The line vortices in the wake induce a downward velocity. This downward velocity is the downwash. In a wing with angle of attack α and free stream velocity V, let the downwash be w. We vectorially add the free stream and the downwash to determine the direction of the effective flow. The effective angle of attack is α_{eff} .

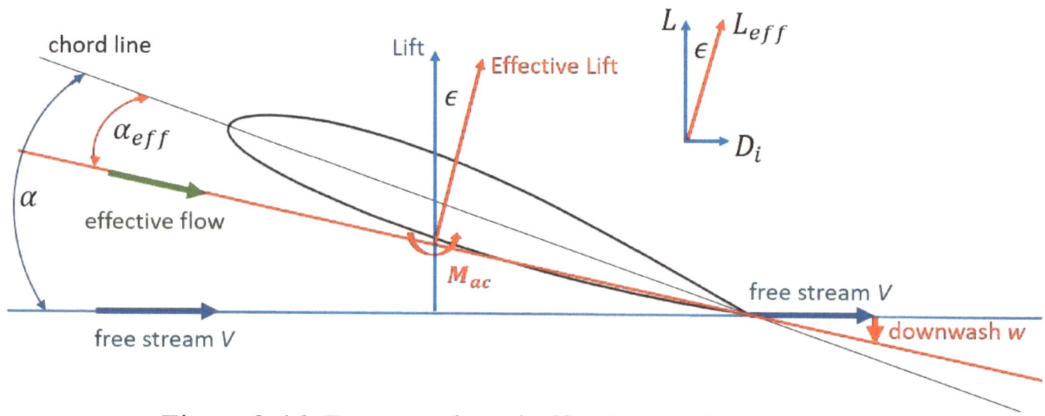

Figure 2.46: Downwash and effective angle of attack.

Lift, by definition, is perpendicular to the free stream. Effective lift is perpendicular to the effective flow. The tilted L_{eff} has a component D_i that is along the free stream direction and this is the induced drag due to downwash. The angle $\epsilon \ll 1$ and

$$L = L_{eff} cos\epsilon \approx L_{eff} \;\; ; \;\; D_i = L_{eff} sin\epsilon \approx L_{eff}\epsilon \approx L\epsilon \tag{2.26}$$

Units of lift and induced drag are *N (lb)* and unit of moment is *N.m (lb.ft)*. We define three unit-less quantities as

$$C_L = \frac{L}{QS} \;\; ; \;\; C_{Di} = \frac{D_i}{QS} = \epsilon C_L \;\; ; \;\; C_{M,ac} = \frac{M_{ac}}{QSc} \;\; ; \;\; Q = \frac{1}{2}\rho V^2 \tag{2.27}$$

These quantities are lift, induced drag, and moment coefficients (note the upper case subscripts), S is the wing area, c is the mean chord, and Q is the dynamic pressure.

Downwash is a consequence of Newton's Third Law. The free stream pushes the wing up and the wing pushes the free stream down. Larger lift will produce larger downwash and larger ϵ. Hence

$$\epsilon \propto C_L \ ; \ \epsilon = KC_L \ ; \ C_{Di} = KC_L^2 \tag{2.28}$$

Analytical solution of wing problems is nearly impossible. Lifting Line Theory and Vortex Lattice Method are fast numerical methods that produce solutions with reasonable accuracy. Commercially available software that employ advanced numerical techniques may take hours to calculate lift and induced drag for a wing.

In the following section, we list a set of semi-empirical relations for quick calculation of lifting properties of a wing. A downloadable Excel Spreadsheet for an eight-vortex, lifting line calculation is available at the URL:

https://drive.google.com/drive/folders/1c_ZpYPUMh_UlKCwwaZ33rwLHAIqCc6sJ?usp=sharing

2.20 Wing

2.20.1 Geometry

Consider a wing with span b , root chord c_R , and tip chord c_T .

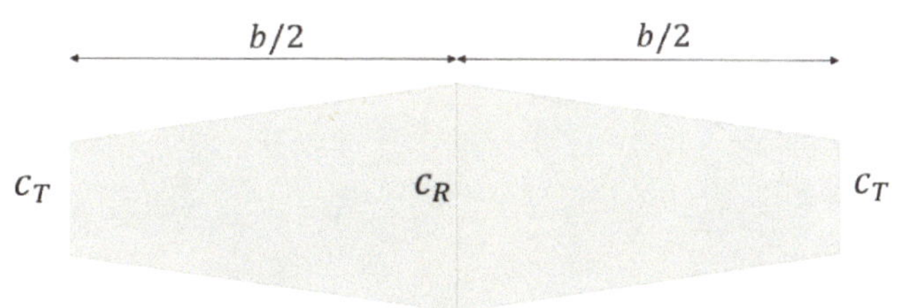

Figure 2.47: Wing geometry.

For taper ratio τ

$$c_T = \tau c_R \tag{2.29}$$

Average chord, wing area, and aspect ratio are

$$c = \frac{1}{2}(c_T + c_R) \ ; \ S = bc \ ; \ AR = \frac{b^2}{S} \tag{2.30}$$

Example-2.10

A wing has an area of 310ft², a wingspan of 54ft, and a taper ratio of 0.7. Determine the aspect ratio, average chord, root chord, and tip chord.

$$AR = \frac{54^2}{310} = 9.41$$

$$c = \frac{310}{54} = 5.74\,ft$$

$$5.74 = 0.5(0.7c_R + c_R)\ ; \quad c_R = 6.75\,ft\ ; \quad c_T = 0.7 \times 6.75 = 4.73\,ft$$

2.20.2 Taper

Consider an untwisted wing with aspect ratio of *8.5*, area of *2500ft²*, and at an angle of attack of *4°*. The airfoil in the wing is NACA 23012. We calculated the lift coefficient by using an eight-vortices lifting line model for three taper ratios.

τ	$c_R\,ft$	$c_T\,ft$	C_L
1.0	17.15	17.15	0.4473
0.6	21.44	12.86	0.4558
0.2	28.58	5.717	0.4573

Figure 2.48: Effect of taper ratio on lift coefficient

In the figure below, we show the lift distribution along the span for the three taper ratios.

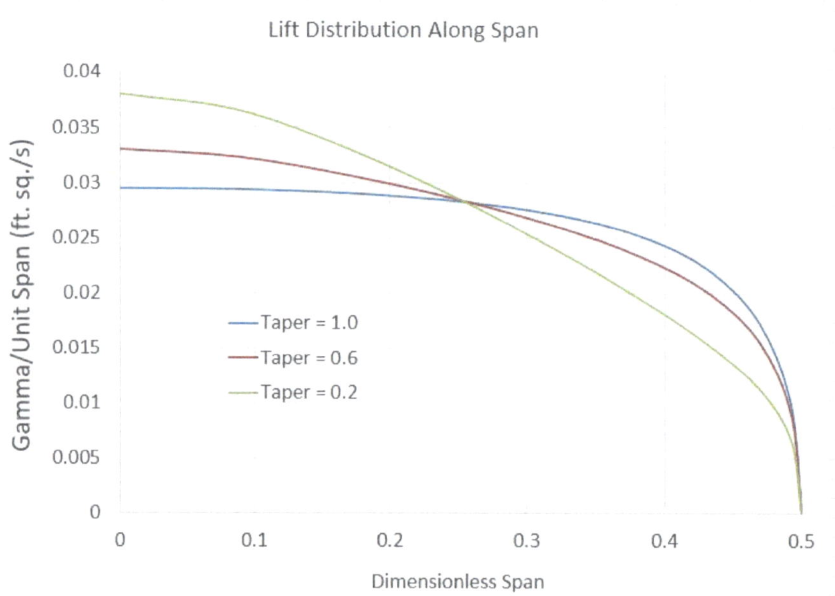

Figure 2.49: Lift distribution along span for three taper ratios.

Observation

- As we increase the taper (decrease the taper ratio), lift coefficient increases.
- As we increase the taper (decrease the taper ratio), the length of the root chord increases and gives larger structural stiffness to the wing.
- As we increase the taper (decrease the taper ratio), the tip produces less lift and the root produces more lift. Smaller lift from the tip produces smaller bending moment at the root where we attach the wing to the fuselage.

Strongly tapered wings have structural and aerodynamic benefits. However, one drawback of taper is that strong taper increases the possibility of tip stall during flight.

2.20.3 Twist

As taper increases, the tip chord decreases and the Reynolds Number based on tip chord-length decreases. The lower Reynolds Number makes the tip stall at a smaller angle of attack (see Figure 2.39). To avoid tip stall, we twist the tip down so that the tip flies at a smaller angle of attack.

Consider a twisted wing with aspect ratio of *8.5*, area of $2500ft^2$, taper ratio of *0.4*, and at an angle of attack of 4°. We calculated the lift coefficient by using an eight-vortex lifting line model for three twist-down angles where the twist-down varies linearly along the span.

Twist (degrees)	C_L
0	0.4586
-1	0.4212
-2	0.3838
-3	0.3464

Figure 2.50: Effect of twist on lift coefficient

In the figure below, we show the lift distribution along the span for the four values of twist.

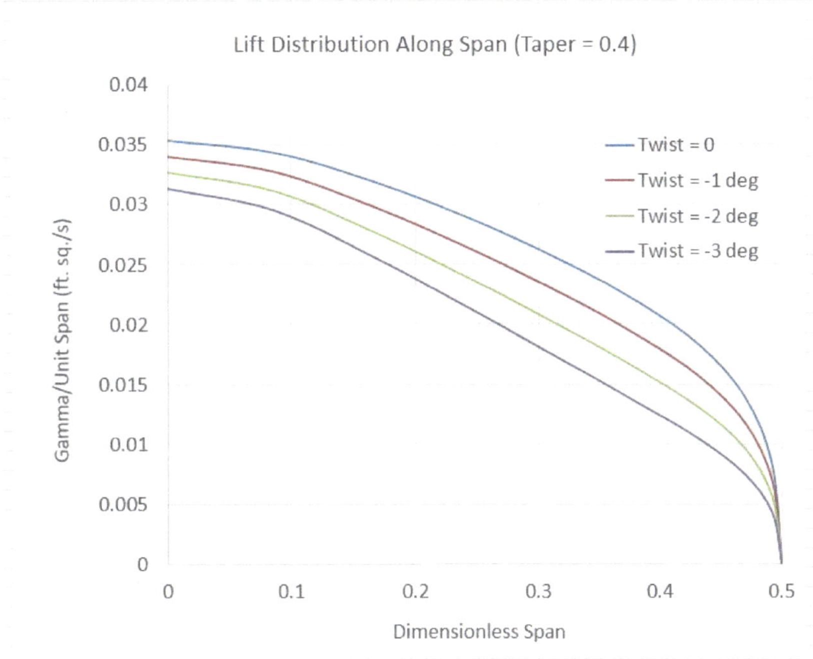

Figure 2.51: Lift distribution along span for four values of twist.

Observation

- A twist down makes the wing lose lift over the entire span.
- Loss of lift increases with the increase in twist down.

A designer's task is to find a wing with desirable lift by combining taper for high lift and twist-down to avoid tip stall. Such design task is beyond the scope of this text; we will continue with untwisted and unswept wings, in the following sections.

2.20.4 Wing Area

During landing, when the airspeed is low, wing area has to be sufficiently large to produce the required lift. The following Figure shows the landing weight, wing area, and wing aspect ratio of a few aircrafts. Typically, the aspect ratio is $7 \leq AR \leq 8.5$.

Aircraft	Landing Weight (lb)	Span (ft)	Wing Area (ft^2)	Aspect Ratio
Airbus 300	309000	147	2800	7.72
Airbus 380	840000	262	9100	7.54
Beechjet 400	15700	44	241	8.03
Boeing 737	103000	93	1100	7.86
Boeing 747	630000	197	5500	7.06
Bombardier CRJ	47000	70	587	8.35
Cessna 750	31800	64	527	7.77
Gulfstream IV	75300	94	1137	7.77

Figure 2.52: Typical wing area and aspect ratio.

The relationship between landing weight and wing area is almost linear as seen from the trend line in the following Figure.

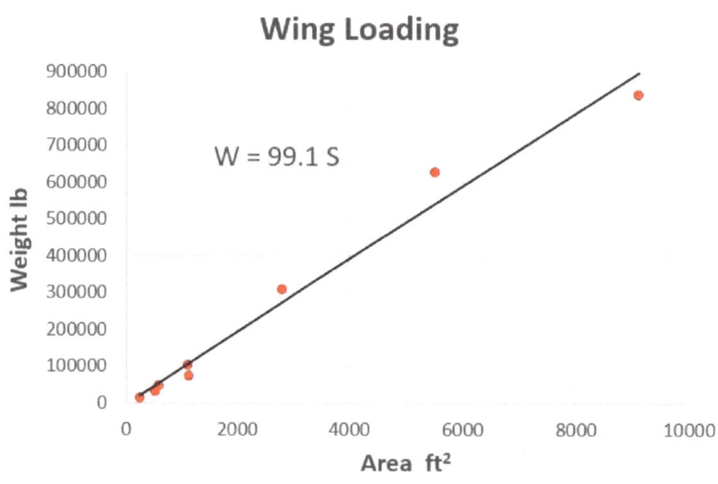

Figure 2.53: Relationship between landing weight and wing area.

2.20.5 Efficiency Factor and Induced Drag Factor

Aspect ratio, efficiency factor, and the induced drag factor are (Cavallo, 1966 and Raymer, 2006, Eqns. 12.48 and 12.49, pg. 347)

$$AR = \frac{b^2}{S} \;;\; e = 1.78(1 - 0.045AR^{0.68}) - 0.64 \;;\; K = \frac{1}{\pi e AR} \tag{2.31}$$

Example-2.11

Determine the efficiency factor and the induced drag factor for aspect ratios 6 and 8.5.

$$e_6 = 1.78 \times (1 - 0.045 \times 6^{0.68}) - 0.64 = 0.8691$$

$$K_6 = \frac{1}{\pi \times 0.8691 \times 6} = 0.06104$$

$$e_{8.5} = 1.78 \times (1 - 0.045 \times 8.5^{0.68}) - 0.64 = 0.7967$$

$$K_6 = \frac{1}{\pi \times 0.7967 \times 8.5} = 0.047$$

Observation

- The induced drag factor decreases as aspect ratio increases.

2.20.6 Lift Coefficient

In the late stages of design, we calculate the lift coefficient of a wing by using elaborate numerical methods. However, at the beginning stages of design, as is the goal in this book, we calculate the lift coefficient for an unswept wing by correcting the slope of the airfoil lift curve a_0 (see Section 2.18) as

$$a = \frac{a_0}{1 + Ka_0} \;;\; C_L = a(\alpha - \alpha_{L=0}) \tag{2.32}$$

To check the accuracy of this formula, we compare the lift coefficient from the formula with the result from eight-vortices lifting line calculation (see Section 2.19).

NACA 23012 Airfoil, 0.6 taper, no twist

Lift Curve Slope	6.188 per rad	0.108 per deg
Zero Lift AOA	-1.3	deg
AOA	4	deg

Aspect Ratio	CL Lifting Line	e	K	CL Formula
6	0.4227	0.8691	0.0610	0.4155
6.5	0.4308	0.8540	0.0573	0.4225
7	0.4381	0.8392	0.0542	0.4287
7.5	0.4446	0.8247	0.0515	0.4342
8	0.4505	0.8106	0.0491	0.4390
8.5	0.4558	0.7967	0.0470	0.4434

Figure 2.54: Comparison between lifting line calculation and Eqn. (2.32).

The comparison shows a discrepancy of about *2%*. The plot shows that the lift coefficient increases almost linearly with aspect ratio in the range *6* to *8.5*.

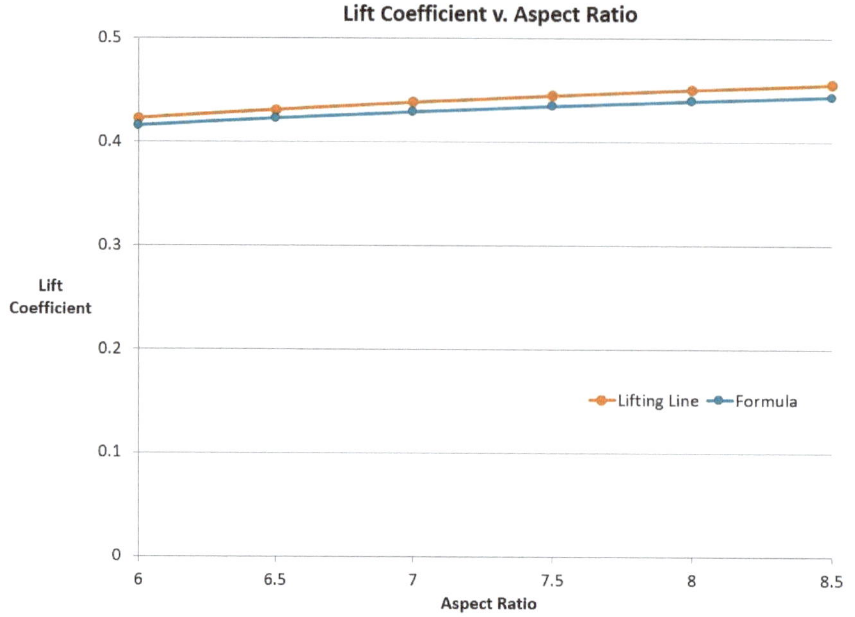

Figure 2.55: Variation of lift coefficient with aspect ratio.

<u>Example-2.12</u>
The airfoil in the wing of an airplane has a lift curve slope of 6.2 rad^{-1} and zero-lift angle of attack of -3°. The area and aspect ratio of the wing are 7.8 and 3000 ft^2. The wing angle of attack is 3°. The wing is flying at an altitude of 10000 ft with 590 ft/s airspeed. Determine the lift force.

From Eqn. (2.31), the span efficiency factor and induced drag factor are

$$e = 1.78 * \{1 - 0.045 * (7.8^{0.68})\} - 0.64 = 0.8162$$

$$K = \frac{1}{\pi \times 0.8162 \times 7.8} = 0.05$$

From Eqn. (2.32), the lift curve slope and the lift coefficient are

$$a = \frac{6.2}{1 + 0.05 \times 6.2} = 4.73 rad^{-1} \;\; ; \;\; C_L = 4.73 \times \left(3 - (-3)\right) \times \frac{\pi}{180} = 0.4953$$

From Eqn. (2.27) and $\rho_{10000} = 0.001756 \frac{slug}{ft^3}$

$$Q = 0.5 \times 0.001756 \times 590^2 = 305.6 \frac{lb}{ft^2}$$

From Eqn. (2.27)

$$L = 0.4953 \times 305.6 \times 3000 = 454000 \; lb$$

2.20.7 Induced Drag Coefficient

To check the accuracy of the induced drag formula (see Equation 2.28), we compare the induced drag coefficient from the formula with the result from an eight-vortices lifting line calculation (see Section 2.19). The comparison shows a discrepancy of about *12%*.

NACA 23012 Airfoil, 0.6 taper, no twist

Lift Curve Slope	6.188 per rad	0.108 per deg
Zero Lift AOA	-1.3	deg
AOA	4	deg

Aspect Ratio	CDi Lifting Line	e	K	CDi Formula
6	0.009638	0.8691	0.0610	0.0105
6.5	0.009260	0.8540	0.0573	0.0102
7	0.008908	0.8392	0.0542	0.0100
7.5	0.008578	0.8247	0.0515	0.0097
8	0.008271	0.8106	0.0491	0.0095
8.5	0.007983	0.7967	0.0470	0.0092

Figure 2.56: Comparison between lifting line calculation and Eqn. (2.28).

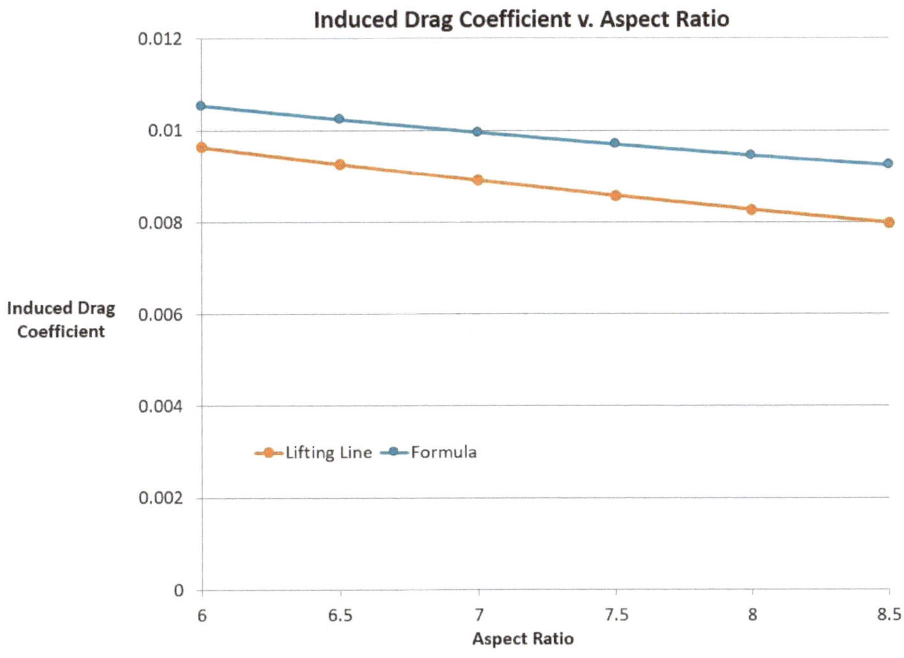

Figure 2.57: Variation of induced drag coefficient with aspect ratio.

The plot shows that the induced drag coefficient decreases almost linearly with aspect ratio. A wing generates lift by producing a low pressure at the top and high pressure at the bottom. The pressure difference drives a span wise flow around the tip and creates wing tip vortices.

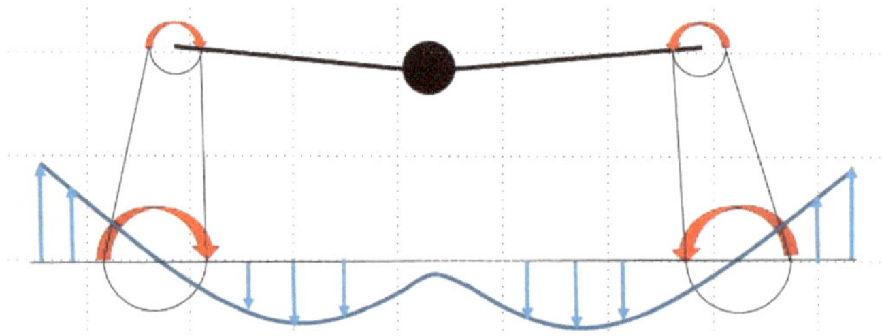

Figure 2.58: Wingtip vortices and downwash

The wing tip vortices cause the downwash and consequently the induced drag. A wing of infinite span (very large aspect ratio) does not have any induced drag because there are no wingtips for the air to leak around. A wing generates less induced drag when the aspect ratio increases and closer it gets to the infinite span approximation.

In unpowered gliders, long slender wings with a high aspect ratio generate less induced drag. Winglets reduce the flow around the wingtip, wingtip vortices, induced drag, and fuel consumption. Double winglets (*15ft* tall with downward extension of *4ft*) on A380 improve fuel efficiency by *4%*.

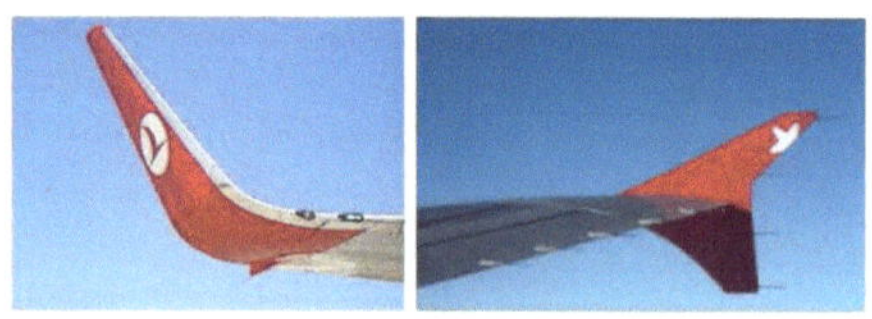

Figure 2.59: Winglets

Example-2.13
Determine the induced drag of the wing of Example-2.12.

From Eqn. (2.26)

$$C_{Di} = 0.05 \times 0.4953^2 = 0.01227$$

From Example-2.12

$$Q = 305.6 \frac{lb}{ft^2}$$

From Eqn. (2.25)

$$D_i = 305.6 \times 3000 \times 0.01227 = 11250\ lb$$

2.20.8 Parasite Drag Coefficient

We determine the parasite drag coefficient of the airplane by using the component buildup method (Raymer, 2006, pg. 328). We will limit ourselves to the wing, the fuselage, and the tail.

$$C_{D0} = C_{D0,wing} + C_{D0,fuselage} + C_{D0,tail} \tag{2.33}$$

The generic formula for the parasite drag coefficient of a component is

$$C_{D0} = \frac{C_f F S_{wet}}{S_{ref}} \tag{2.34}$$

1. Wing and Tail:
    a. We find Re and C_f from Eqn. (2.17) by using the average chord length of the wing or tail in the Reynolds number formula.
    b. The form factor F depends on the thickness to chord ratio of the airfoil in the wing or tail. The thickness to chord ratio is the last two digits in the numbering convention of NACA airfoils. For example, NACA 1624 has a thickness to chord ratio of *0.24*, and NACA 23012 has a thickness to chord ratio of *0.12*.
    c. S_{wet} is double the wing or tail area and S_{ref} is the wing area

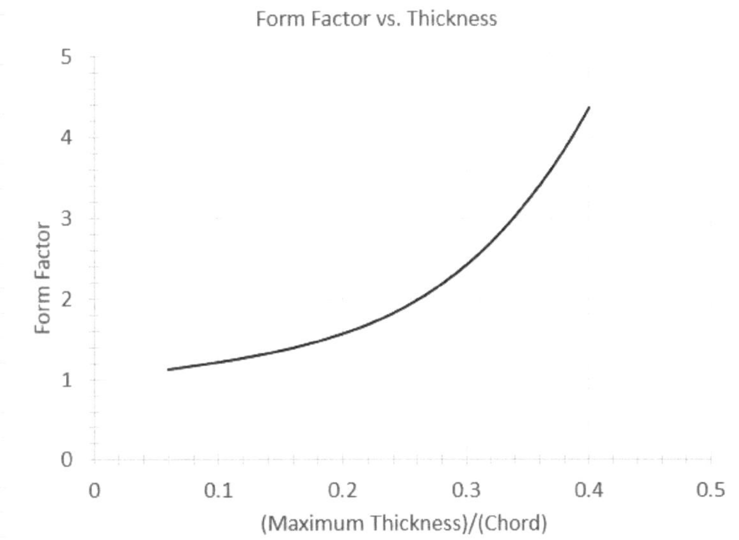

Figure 2.60: Form factor of wing and tail.

2. Fuselage:
    a. We find Re and C_f from Eqn. (2.17) by using the length of the fuselage (ℓ) in the Reynolds number formula.
    b. We find the form factor F from

$$A_{max} = Maximum\ cross\ sectional\ area\ of\ the\ fuselage$$

$$f = \frac{\ell}{\sqrt{(4/\pi)A_{max}}} \ ; \ F = 1 + \frac{60}{f^3} + \frac{f}{400} \tag{2.35}$$

    c. S_{wet} is the fuselage surface area (see Appendix A) and S_{ref} is the wing area

Example-2.14

The airfoil in a wing is NACA 1224. The average chord and the area of the wing are 15 ft and 3000 ft². For an airspeed of 600 ft/s, determine the parasite drag coefficient.

We take (see Section 2.13)

$$C_f = 0.00275$$

From Figure-2.51, for thickness to chord ratio of 0.24 for NACA 1224, form factor F=1.8.

$$S_{wet} = 2 \times 3000 = 6000\ ft^2 \quad ; \quad S_{ref} = 3000\ ft^2$$

$$C_{D0,wing} = \frac{0.00275 \times 1.8 \times 6000}{3000} = 0.0099$$

Example-2.15

In an airplane, the fuselage is 150 ft long, has a maximum cross-sectional area of 140ft², and a surface area of 6600 ft². The wing area is 3000ft². For an airspeed of 600 ft/s, determine the parasite drag coefficient.

We take (see Section 2.13)

$$C_f = 0.00275$$

$$f = \frac{150}{\sqrt{(4/\pi)140}} = 11.23\,; \quad F = 1 + \frac{60}{11.23^3} + \frac{11.23}{400} = 1.0704$$

$$S_{wet} = 6600\ ft^2 \quad ; \quad S_{ref} = 3000\ ft^2$$

$$C_{D0,fuselage} = \frac{0.00275 \times 1.0704 \times 6600}{3000} = 0.0065$$

The total drag coefficient is the sum of induced drag coefficient and the parasite drag coefficient (see Section 2.13).

$$\frac{D}{\frac{1}{2}\rho V^2 S} = C_D = C_{Di} + C_{D0} = K C_L^2 + C_{D0} \tag{2.36}$$

2.20.9 Moment Coefficient

We calculate the wing pitching moment by adding an aspect ratio correction to the airfoil pitching moment (Raymer, 2006, Eqn. 16.19, pg. 479). For an unswept wing

$$C_{M,ac} = C_{m,ac} \frac{AR}{AR + 2} \tag{2.37}$$

$C_{m,ac}$ is the airfoil moment coefficient at the aerodynamic center (see Section 2.17).

Example-2.16
The moment coefficient at the aerodynamic center of an airfoil in the wing of an airplane is -0.02. The area and aspect ratio of the wing are 7.8 and 3000 ft². The mean chord of the wing is 15ft. The wing is flying at an altitude of 10000 ft with 590 ft/s airspeed. Determine the moment at the aerodynamic center of the wing.

From Eqn. (2.37),

$$C_{M,ac} = -0.02 \times \frac{7.8}{7.8 + 2} = -0.01592$$

From Example-2.12

$$Q = 305.6 \frac{lb}{ft^2}$$

From Eqn. (2.27)

$$M_{ac} = -0.01592 \times 305.6 \times 3000 \times 15 = -218900\ lb.ft$$

2.21 Drag Polar

We can calculate the lift and drag coefficients for a set of angles of attack in the operating range. When we plot the lift and drag coefficients, we get the drag polar.

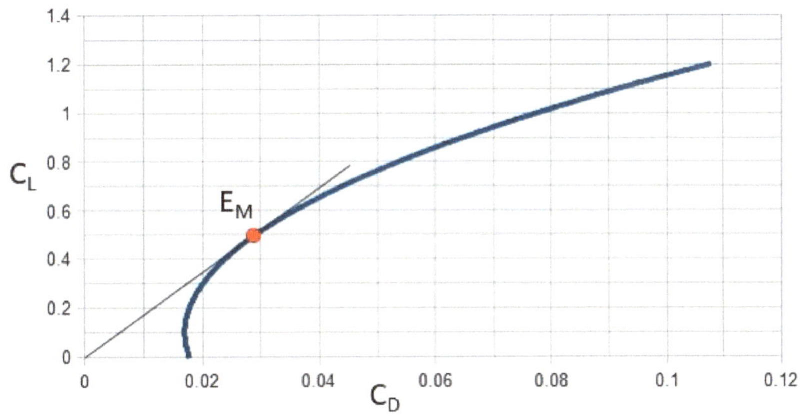

Figure 2.61: Drag polar.

The lift-drag combination marked by E_M corresponds to maximum lift to drag ratio. To maximize the ratio, we set

$$\frac{d\left(\frac{C_L}{C_D}\right)}{dC_L} = \frac{d}{dC_L}\left[\frac{C_L}{C_{D0} + KC_L^2}\right] = 0$$

After simplification, at E_M

$$C_{Di,EM} = C_{D0} \tag{2.38}$$

$$C_{D,EM} = 2C_{D0} \tag{2.39}$$

$$C_{L,EM} = \sqrt{\frac{C_{D0}}{K}} \tag{2.40}$$

$$E_M = \frac{C_{L,EM}}{C_{D,EM}} = \sqrt{\frac{1}{4KC_{D0}}} \tag{2.41}$$

<u>Example-2.17</u>

The wing aspect ratio is 8 and $C_{D0} = 0.0176$. Determine E_M and the lift and drag coefficients. The lift curve slope of the airfoil is 0.1 deg^{-1} and the zero lift angle of attack is -3°. Determine the angle of attack of the wing.

The efficiency factor is

$$e = 1.78 * \{1 - 0.045 * (8^{0.68})\} - 0.64 = 0.8106$$

The induced drag coefficient is

$$K = \frac{1}{\pi \times 0.8106 \times 8} = 0.0491$$

From Eqn. (2.40), the lift coefficient at E_M is

$$C_{L,EM} = \sqrt{\frac{0.0176}{0.0491}} = 0.5987$$

From Eqn. (2.39), the drag coefficient at E_M is

$$C_{D,EM} = 2 \times 0.0176 = 0.0352$$

The lift-drag ratio at E_M is

$$\left(\frac{C_L}{C_D}\right)_{EM} = 17$$

From Eqn. (2.32), lift curve slope of the wing is

$$a = \frac{0.1}{1 + 0.0491 \times .1} = 0.0995\ deg^{-1}$$

From Eqn. (2.32), the angle of attack of the wing is

$$0.5987 = 0.0995\{\alpha - (-3)\}\ ; \quad \alpha = 3.017^o$$

2.22 Stall Airspeed and Maximum Lift Coefficient

During level flight, the lift must balance the weight of the airplane.

$$W = L = \frac{1}{2}\rho V^2 S C_L \ ; \quad V = \sqrt{\frac{2\left(\frac{W}{S}\right)}{\rho C_L}}$$

The quantity $\frac{W}{S}$ is the wing loading that has a typical value of around *100 lb/ft²*. For a fixed altitude, an airplane can fly at a lower speed by increasing the angle of attack that increases the lift coefficient. However, the upper bound of the angle of attack is the stall angle of attack. The lowest speed, V_{stall} of the airplane corresponds to the maximum lift coefficient at stall. The stall airspeed increases with the increase in altitude, as density decreases with increase in altitude.

$$V_{stall} = \sqrt{\frac{2\left(\frac{W}{S}\right)}{\rho C_{L,Max}}} \tag{2.42}$$

<u>*Example-2.18*</u>

In an airplane, the wing loading is 100 lb/ft² and maximum lift coefficient is 0.61. Determine the stall airspeed at sea level and at an altitude of 30,000ft.

$$\rho_{SL} = 0.00238\frac{slug}{ft^3}; V_{stall,SL} = \sqrt{\frac{2 \times 100}{0.00238 \times 0.61}} = 371.2\frac{ft}{s} = 253.1\frac{mi}{h}$$

$$\rho_{30000} = 0.000891\frac{slug}{ft^3}; V_{stall,30000} = \sqrt{\frac{2 \times 100}{0.000891 \times 0.61}} = 606.6\frac{ft}{s} = 413.6\frac{mi}{h}$$

The stall airspeed is smaller at lower altitude.

2.23 Flaps

During landing, we may need to fly the airplane at a speed lower than the stall speed of Eqn. (2.42). In such a situation, the pilot increases the maximum lift coefficient by deploying the flap that is a high lift device.

Figure 2.62: Flaps

There are several types of flaps (Raymer, 2006, pg. 323)

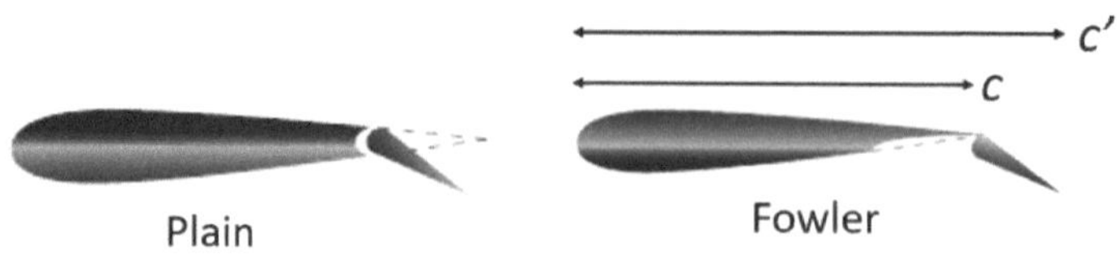

Figure 2.63: Types of flaps.

The flap increases the lift by (Raymer, 2006, pg. 326)

$$\Delta C_{L,Max} = 0.9\Delta C_{l,Max}\left(\frac{S_{flapped}}{S}\right)\cos\Lambda_{HL} \tag{2.43}$$

Approximate lift contributions of high lift devices (Raymer, 2006, pg. 326) are

Device	$\boldsymbol{\Delta C_{l,Max}}$
Plain and split	0.9
Slotted	1.3
Fowler	$1.3\left(\frac{c'}{c}\right)$
Double slotted	$1.6\left(\frac{c'}{c}\right)$
Triple slotted	$1.9\left(\frac{c'}{c}\right)$

Figure 2.64: Lift gain from flap

Flap settings for takeoff and landing are different. We use 60-80% of the value from Eqn. (2.43) for takeoff (Raymer, 2006, pg. 326)

$$\Delta C_{L,Max,TO} = 0.7\Delta C_{L,Max}\ ;\ \Delta C_{L,Max,LA} = \Delta C_{L,Max} \tag{2.44}$$

The zero-lift angle of attack decreases with the deployment of the flaps.

$$\Delta\alpha_{L=0} = \Delta\alpha_{L=0,2D}\left(\frac{S_{flapped}}{S}\right)\cos\Lambda_{HL} \tag{2.45}$$

Takeoff	$\Delta\alpha_{L=0,2D} = -10^0$
Landing	$\Delta\alpha_{L=0,2D} = -15^0$

Figure 2.65: Change in zero-lift angle of attack due to flap.

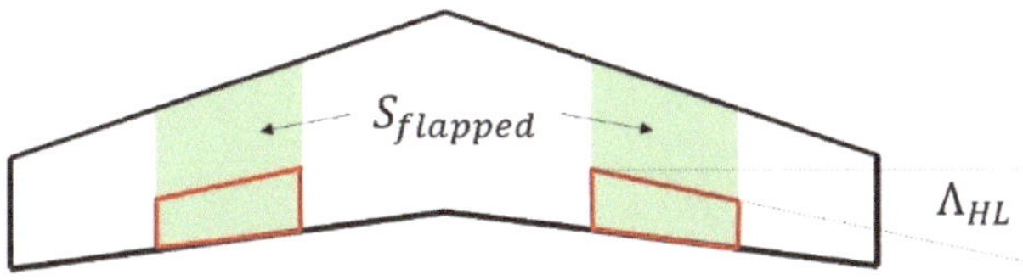

Figure 2.66: Flap geometry

The sweepback angle of the hinge line of the flap is Λ_{HL} , and the wing area is S . Flaps lower the airplane stalling speed by increasing the lift coefficient (see Eqn. 2.42) and the airplane can land and takeoff at a lower speed.

We calculate the increase in the induced drag due to the increase in lift coefficient from (Raymer, 2006, Eqn. 12.62)

$$\Delta C_{Di} = k_f^2 (\Delta C_{L,Max})^2 \cos \Lambda_{HL} \tag{2.46}$$

The factor k_f is *0.14* for full span flap and *0.28* for half-span flaps.

<u>Example-2.19</u>
The airfoil in an unswept wing has a lift-curve slope of 0.1 per degree and a zero-lift angle of attack is -3°. The induced drag factor K = 0.05. In a Fowler flap, the ratio of the flapped area and the wing area is 0.28, the ratio of chords with and without flap is 1.2. Determine the maximum lift coefficient for takeoff when the wing angle of attack is 4°.

From Figure 2.64,

$$\Delta C_{l,Max} = 1.3 \times 1.2 = 1.56$$

From Eqn. (2.43)

$$\Delta C_{L,Max} = 0.9 \times 1.56 \times 0.28 = 0.3931$$

From Eqn. (2.44),

$$\Delta C_{L,Max,TO} = 0.7 \times 0.3931 = 0.2752$$

From Eqn. (2.32),

$$a = \frac{0.1}{1 + 0.05 \times .1} = 0.09950$$

From Eqn. (2.45) and Figure 2.65,

$$\Delta \alpha_{L=0} = -10 \times 0.28 = -2.8\, deg$$

$$\alpha_{L=0} = -3 - 2.8 = -5.8\, deg$$

From Eqn. (2.32),

$$C_L = 0.09950 \times (4 + 5.8) = 0.9751$$

$$C_{L,Max,TO} = C_L + \Delta C_{L,Max,TO} = 0.9751 + 0.2752 = 1.250$$

Example-2.20
The airfoil in an unswept wing has a lift-curve slope of 0.1 per degree and a zero-lift angle of attack is -3°. The induced drag factor K = 0.05. In a Fowler flap, the ratio of the flapped area and the wing area is 0.28, the ratio of chords with and without flap is 1.2. Determine the maximum lift coefficient for landing when the wing angle of attack is 6°.

From Figure 2.64,

$$\Delta C_{l,Max} = 1.3 \times 1.2 = 1.56$$

From Eqn. (2.43)

$$\Delta C_{L,Max} = 0.9 \times 1.56 \times 0.28 = 0.3931$$

From Eqn. (2.44),

$$\Delta C_{L,Max,LA} = 0.3931$$

From Eqn. (2.32),

$$a = \frac{0.1}{1 + 0.05 \times .1} = 0.09950$$

From Eqn. (2.45) and Figure 2.60,

$$\Delta \alpha_{L=0} = -15 \times 0.28 = -4.2\ deg$$

$$\alpha_{L=0} = -3 - 4.2 = -7.2\ deg$$

From Eqn. (2.32),

$$C_L = 0.09950 \times (6 + 7.2) = 1.313$$

$$C_{L,Max,LA} = C_L + \Delta C_{L,Max,LA} = 1.313 + 0.3931 = 1.707$$

2.24 V-n Diagram

2.24.1 Definition

During takeoff and landing, an airplane flies on a curved path on the vertical plane. During a banked turn, the airplane flies on a curved path on a horizontal plane. During these flights

$$L > W$$

The load factor is the ratio of lift and weight.

$$n = \frac{L}{W}$$

For a curved path on the vertical plane, the larger lift causes the centripetal acceleration that the airplane needs to fly on a curved path.

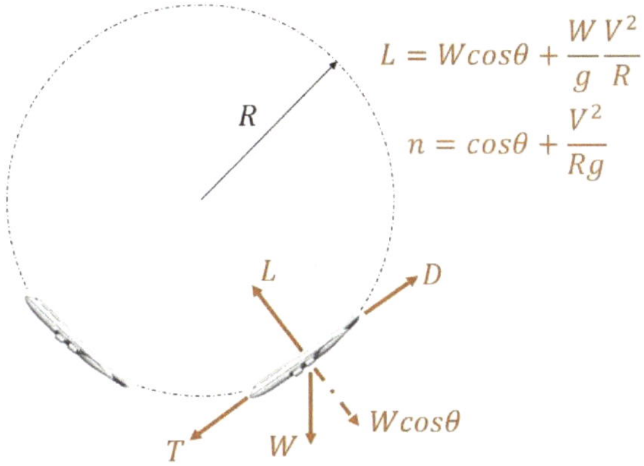

Figure 2.66: Curved path on a vertical plane.

For light airplanes, the load factor just before ground contact is less than *1.2* (Hall, 1970). For a curved path on the horizontal plane, during a bank turn

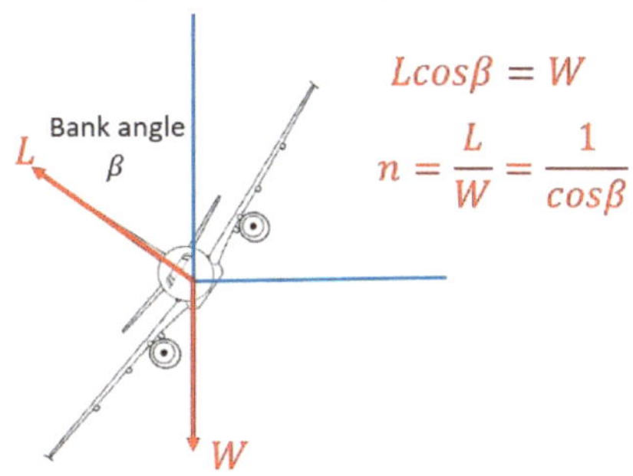

Figure 2.67: Curved path on a horizontal plane.

During a bank turn, we determine the upper limit of load factor by considering stall and structural integrity. For detail discussion on bank turn, see Chapter 6.

$$n = \frac{L}{W} = \frac{\rho V^2 C_L}{2\left(\frac{W}{S}\right)} \tag{2.47}$$

When we insert the maximum (positive) and the minimum (negative) values of the lift coefficient in Eqn. (2.47), we find the upper and lower bounds of the load factor that arise from stall at a fixed speed.

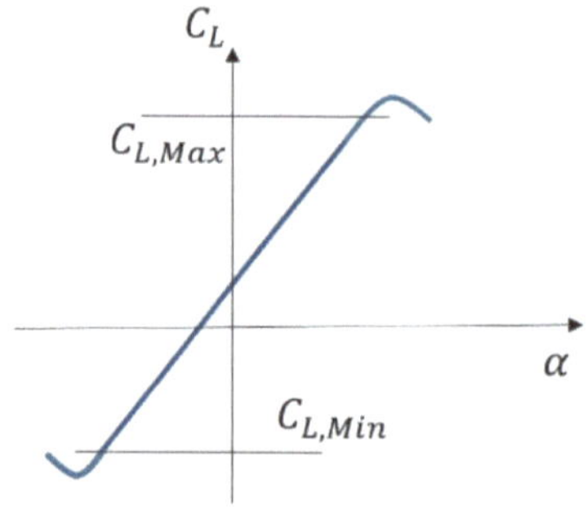

Figure 2.68: Maximum and minimum left coefficient.

$$n_{stall+} = \frac{\rho V_{stall}^2 C_{L,Max}}{2\left(\frac{W}{S}\right)} \quad ; \quad n_{stall-} = \frac{\rho V_{stall}^2 C_{L,Min}}{2\left(\frac{W}{S}\right)} \qquad (2.48)$$

<u>*Example-2.21*</u>

In an airplane, the wing loading is 110lb/ft² and maximum lift coefficient is 0.6. The airplane performs a maneuver at an altitude of 10,000ft with upper bound of load factor as 1.8. Determine the stall airspeed.

$$\rho_{10000} = 0.001767 \frac{slug}{ft^3}$$

$$1.8 = \frac{0.001767 \times V_{stall}^2 \times 0.6}{2 \times 110} \quad ; \quad V_{stall} = 611.2 \frac{ft}{s} = 416.7 \frac{mi}{h}$$

Load factor is important for two reasons:

- High load factor imposes overload on the airplane structure.
- Increased load factor increases the stalling speed and the airplane may stall at seemingly safe flight speeds.

In the *V-n* diagram, we plot speed along the abscissa, load factor along the ordinate, and show the flight envelope of an airplane that is limited by stall and structural integrity.

<u>*2.24.2 Drawing the V-n Diagram*</u>

- The V-n Diagram depends on the altitude because density is a function of altitude.
- We must know the wing loading and the maximum lift coefficient of the airplane.
- We choose a value of the load factor and determine the corresponding stall airspeed from Eqn. (2.45). The (*stall airspeed, load factor)* pair is one point on the stall curve (see *Example-2.21)*. By plotting several such points, we can draw the stall curve.
- Load factor is one for level flight.

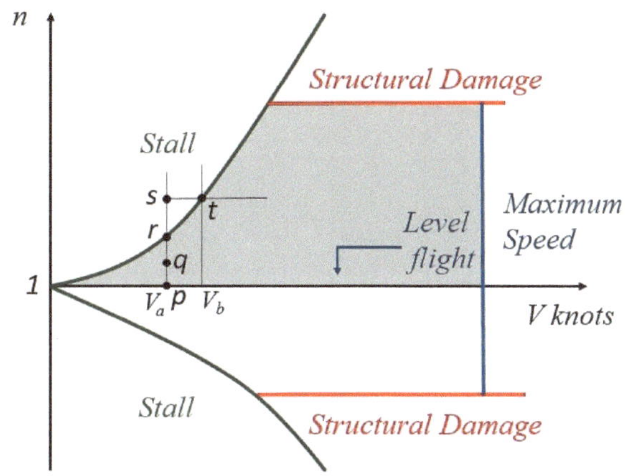

Figure 2.69: V-n diagram

2.24.3 Interpretation of V-n Diagram

For a fixed airspeed V_a

- At flight condition p, the angle of attack is such that the lift coefficient $C_{L,p}$ is sufficient for a level flight. Load factor is one and lift equals weight.
- At flight condition q, the airplane is performing a maneuver with load factor more than one and the angle of attack is such that the lift coefficient
$$C_{L,p} < C_{L,q} < C_{L,Max}$$
- At flight condition r, the airplane is performing a maneuver with load factor on the stall boundary and the angle of attack is such that the lift coefficient
$$C_{L,q} < C_{L,r} = C_{L,Max}$$
- At flight condition s, the airplane is performing a maneuver with load factor beyond the stall boundary and the angle of attack is such that the lift coefficient
$$C_{L,s} > C_{L,Max}$$
The airplane has stalled. The pilot can take two possible actions – reduce the angle of attack to bring the operating point to r or increase the airspeed to V_b to bring the operating point to t.

The two load factors n_{stall+} and n_{stall-} limit the flight envelop on the left side. At the top and bottom, the magnitude of the load factor and the lift are too large to cause structural damage in the airplane. At the right, the speed is too high and can cause structural damage in the airplane.

2.25 Dihedral and Sweepback

In addition to aspect ratio, taper, and twist, the wing has two more important, geometric characteristics – dihedral and sweepback.

2.25.1 Dihedral

Dihedral is the upwardly turned angle of an airplane's wings that gives roll stability to the airplane.

Figure 2.70: Dihedral angle

Consider a wing with dihedral γ that has a roll angle θ_R . The roll tilts the lift force that has a horizontal component L_H . This component makes the airplane slip to the left with a slip velocity V_{Slip} . The relative airspeed due to the slip is $V_{Slip,rel}$ to the right.

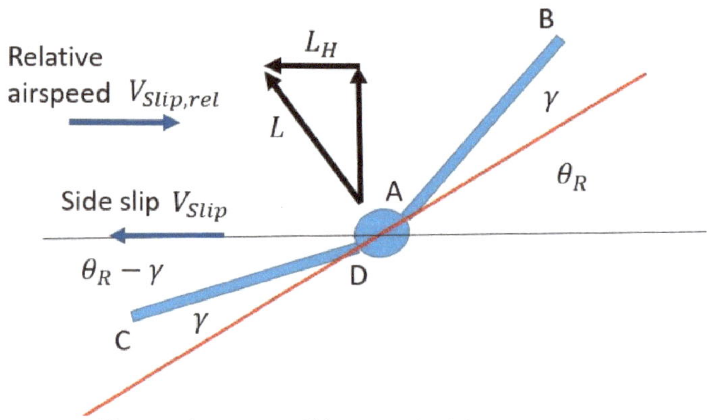

Figure 2.71: Rolling and side slip

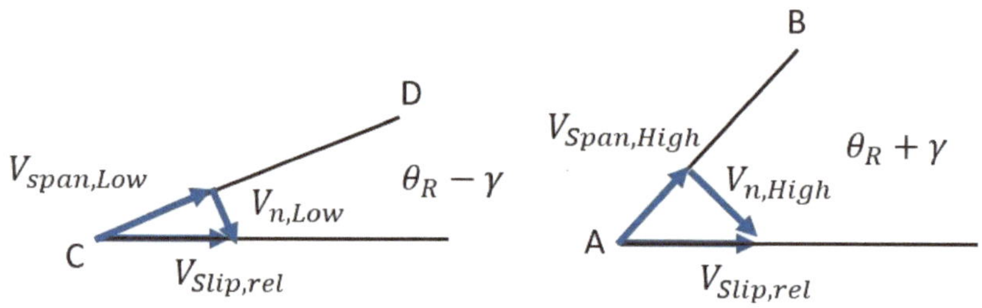

Figure 2.72: Downwash due to dihedral

For the high half-wing *AB*, we resolve the relative airspeed due to slip into two components. One component, $V_{span,High}$, is along the span of the high half-wing *AB* (*along line AB)* that has no effect on the lift. The other component, $V_{n,High}$, is a downwash perpendicular to the high half-wing *AB* (*perpendicular to line AB*).

$$V_{n,High} = V_{Slip,rel} \sin(\theta_R + \gamma) \tag{2.49}$$

Similarly, for the low half-wing *CD*, we resolve the relative airspeed due to slip into two components. One component, $V_{span,Low}$, is along the span of the low half-wing *CD* (*along line CD*) that has no effect on the lift. The other component, $V_{n,Low}$, is a downwash perpendicular to the low half-wing *CD* (*perpendicular to line CD*).

$$V_{n,Low} = V_{Slip,rel} \sin(\theta_R - \gamma) \tag{2.50}$$

By comparing Eqns. (2.49) and (2.50), the downwash on the low half-wing is smaller than the downwash on the high half-wing.

$$V_{n.Low} < V_{n,High}$$

Note that without the dihedral, $\gamma = 0$, the downwash on the high and the low half-wings are equal. The smaller downwash on the low half-wing gives it a larger angle of attack compared to the high half-wing that has a larger downwash. The larger angle of attack for the low half-wing causes a larger lift from the low half-wing compared to the lift from high half-wing.

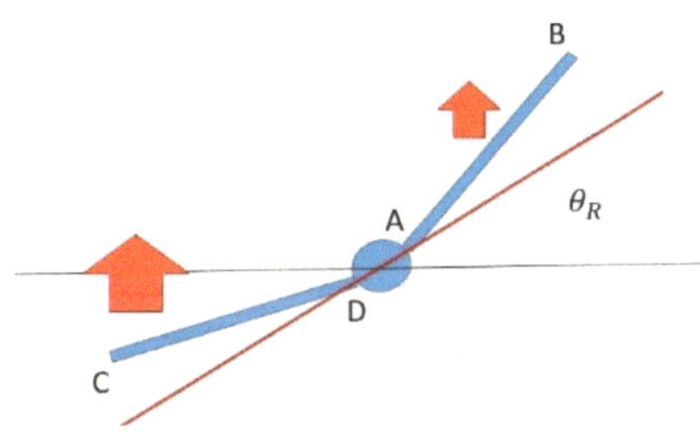

Figure 2.72: Restoring moment due to dihedral

The larger lift from the low half-wing restores the airplane to its level orientation from its rolled orientation.

2.25.2 Sweepback

The sweepback prevents certain events from occurring on the upper surface of the wing due to compressibility of air at high speeds and that cause reduction in lift and enhanced drag. Such effects of compressibility appear when the airspeed, at any point on the top surface, approaches or exceeds the local speed of sound. For convenience in monitoring these effects, we define Mach number as the ratio of the local airspeed and the speed of sound.

$$M = \frac{local\ airspeed}{speed\ of\ sound}$$

- $M < 1$; the flow is subsonic
- $M = 1$; the flow is sonic
- $M > 1$; the flow is supersonic

By examining a section of the wing, the velocity of air increases as it passes over the hump at the top surface of the wing. As the subsonic Mach of the airspeed increases and reaches a critical value, at the location of the highest velocity on the top surface, the flow becomes sonic.

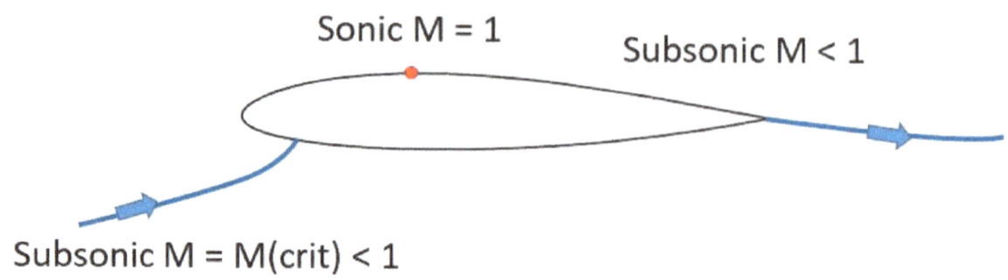

Figure 2.73: Critical Mach number

When the Mach of the airspeed is larger than the critical Mach, the flow in the front portion of the top surface becomes supersonic. This supersonic flow passes through a shock and becomes subsonic in the rear portion of the top surface.

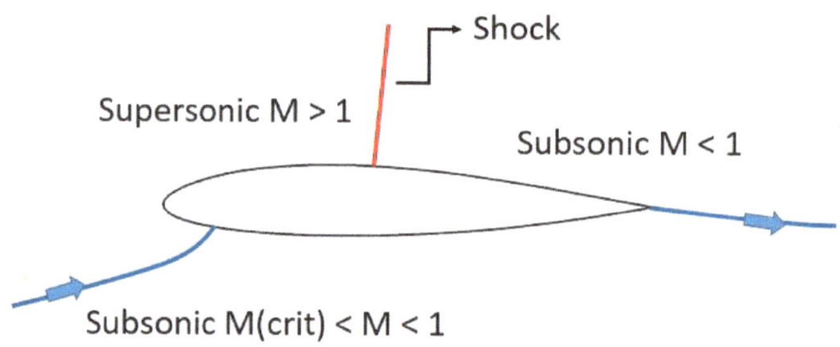

Figure 2.74: Shock at the top surface

The main characteristics of a shock are:

- Abrupt rise of pressure across a shock
- Abrupt drop of velocity across a shock

An illustrative variation of pressure as we travel along the top surface of the wing-section from the front stagnation point to the trailing edge is

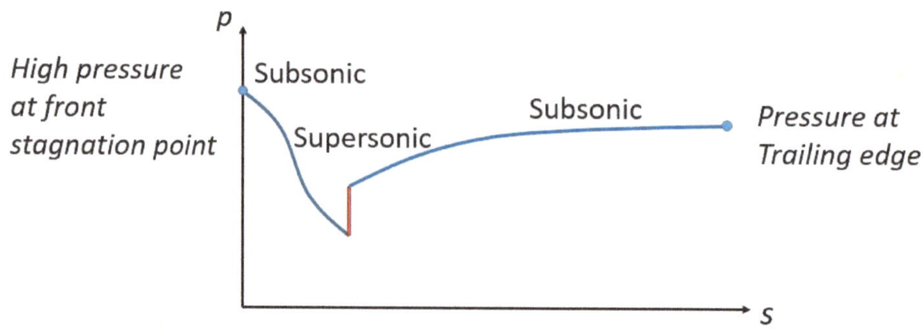

Figure 2.75: Variation of pressure along the top surface

Observation

- Pressure is the highest at the front stagnation point where the velocity is zero
- Pressure decreases as the velocity increases to sonic from subsonic
- Pressure continues to decrease as the velocity becomes supersonic
- Abrupt rise in pressure and decrease in velocity across the shock to subsonic
- Pressure rises and velocity decreases from the shock to the trailing edge

The presence of the shock increases the drag of the wing. A discussion on drag rise Mach number is beyond the scope of this book.

Furthermore, the abrupt pressure rise across the shock causes the flow to separate downstream from the shock (see Section 2.16).

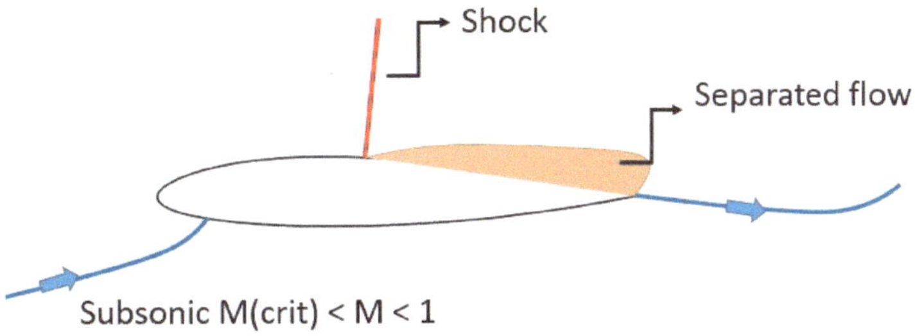

Figure 2.76: Separation after shock

The separated flow downstream from the shock reduces the lift of the wing.

When we sweep the wing, the "effective Mach" seen by the wing is the component of the Mach that is perpendicular to the leading edge of the wing.

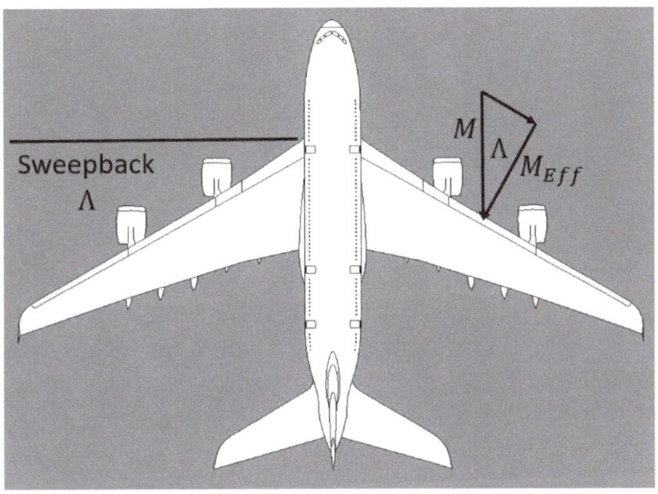

Figure 2.76: Sweepback and effective Mach

$$M = \frac{M_{Eff}}{cos\Lambda}$$

$$M_{Crit} = \frac{M_{Eff,Crit}}{cos\Lambda}$$

For a wing with critical Mach of *0.7* and a sweep angle of *20º*, the critical Mach for flight is

$$M_{Crit} = \frac{0.7}{cos20^o} = 0.7449$$

The *20º* sweepback enables the wing to fly at a Mach of *0.7449* (instead of *0.7*) without any loss of lift or increase in drag due to the compressibility effects in the flow.

3 Cruise

3.1 Equations of Motion

During cruise, an airplane flies along a horizontal path with a uniform airspeed. As the airplane is not climbing or descending, the vertical forces on the airplane are balanced. As the airplane is not accelerating, the horizontal forces are balanced.

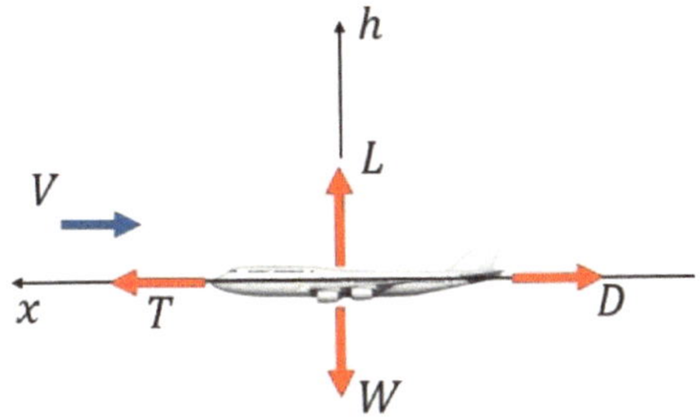

Figure 3.1: Force balance in cruise.

$$L = W \tag{3.1}$$

$$D = T \tag{3.2}$$

The equations of kinematics for zero acceleration are

$$\frac{dx}{dt} = V \tag{3.3}$$

$$\frac{dh}{dt} = 0 \tag{3.4}$$

$$\frac{dV}{dt} = 0 \tag{3.5}$$

3.2 Thrust versus Airspeed

With ρ being the density at the cruise altitude, S being the wing area, and C_L being the lift coefficient, Eqn. (3.1) becomes

$$W = \frac{1}{2}\rho V^2 S C_L$$

Or

$$C_L = \frac{2(W/S)}{\rho V^2} \tag{3.6}$$

By using Eqns. (2.28) and (2.36),

$$C_D = C_{DO} + KC_L^2 = C_{DO} + K\frac{4\left(W/_S\right)^2}{\rho^2 V^4}$$

By using Eqn. (2.36) and Eqn. (3.2),

$$T = D = \frac{1}{2}\rho V^2 S C_D = \frac{1}{2}\rho V^2 S C_{DO} + K\frac{2S\left(W/_S\right)^2}{\rho V^2} \tag{3.7}$$

Or

$$\rho^2 V^4 C_{DO} - 2\left(T/_S\right)\rho V^2 + 4K\left(W/_S\right)^2 = 0 \tag{3.8}$$

By solving the quadratic equation (3.8) for V^2,

$$V^2 = \frac{2\left(T/_S\right)\rho \pm \sqrt{4\left(T/_S\right)^2 \rho^2 - 16K\rho^2 C_{DO}\left(W/_S\right)^2}}{2\rho^2 C_{DO}}$$

By simplifying,

$$V^2 = \frac{\left(T/_S\right) \pm \sqrt{\left(T/_S\right)^2 - 4KC_{DO}\left(W/_S\right)^2}}{\rho C_{DO}} \tag{3.9}$$

From Eqn. (2.41),

$$4KC_{DO} = \frac{1}{E_M^2} \tag{3.10}$$

K is the induced drag coefficient (see Eqn. 2.31)). By combining Eqns. (3.9) and (3.10),

$$V^2 = \frac{\left(T/_S\right) \pm \sqrt{\left(T/_S\right)^2 - \frac{1}{E_M^2}\left(W/_S\right)^2}}{\rho C_{DO}} \tag{3.11a}$$

An alternative form of Eqn. (3.11a) is

$$V = \left[\frac{\left(\frac{T}{S}\right)}{\rho C_{DO}}\left[1 \pm \sqrt{1 - \left\{\frac{1}{E_M\left(\frac{T}{W}\right)}\right\}^2}\right]\right]^{\frac{1}{2}} \tag{3.11b}$$

3.2.1 Level Flight not Possible

When the quantity under the radical sign, in Eqn. (3.11a), is negative, level flight is not possible.

$$\frac{T}{S} < \frac{1}{E_M}\frac{W}{S} \quad ; \quad \frac{T}{W} < \frac{1}{E_M} \tag{3.12}$$

3.2.2 Level Flight with Minimum Thrust

When the quantity under the radical sign, in Eqn. (3.11a), is zero, thrust is minimum for level flight.

$$\frac{T}{S} = \frac{1}{E_M}\frac{W}{S} \quad ; \quad \frac{T}{W} = \frac{1}{E_M} \tag{3.13}$$

Example 3.1

A 150000lb airplane has $E_M = 16$. Determine the minimum thrust for level flight.

From Eqn. (3.13),

$$T = \frac{150000}{16} = 9375\ lb$$

Note: E_M depends on the induced drag coefficient of the wing and C_{D0}, but not on density. Hence, the minimum thrust does not depend on the altitude. However, from Eqn. (3.11), the airspeed at minimum thrust depends on density (altitude).

Example 3.2

A 150000lb airplane has $E_M = 16$, $C_{D0} = 0.02$, and wing area $S = 1600ft^2$. Determine the airspeed at minimum thrust at 10000ft altitude ($\rho = 0.001756slug/ft^3$).

From Eqn. (3.13),

$$T = \frac{150000}{16} = 9375lb \ ; \ \frac{T}{S} = \frac{9375}{1600} = \frac{5.859ft}{s}$$

From Eqn. (3.11),

$$V = \sqrt{\frac{5.895}{0.001756 \times 0.02}} = 408.5\frac{ft}{s}$$

3.2.3 Level Flight with More than Minimum Thrust

When the thrust is more than the minimum, the quantity under the radical sign, in Eqn. (3.11a), is positive; there are two possible velocities for level flight. One corresponding to the + sign in front of the radical and the other corresponding to the $-$ sign in front of the radical.

$$\frac{T}{W} > \frac{1}{E_M} \tag{3.14}$$

<u>*Example 3.3*</u>

A 150000lb airplane has $E_M = 16$, $C_{D0} = 0.02$, and wing area $S = 1600ft^2$. Determine the airspeed at 1.2 times the minimum thrust at 10000ft altitude ($\rho = 0.001756slug/ft^3$).

From Eqn. (3.13),

$$T = \frac{150000}{16} \times 1.2 = 11250\ lb$$

$$\frac{T}{S} = \frac{11250}{1600} = 7.031\frac{lb}{ft^2}\ ;\ \ \frac{W}{S} = \frac{150000}{1600} = 93.75\frac{lb}{ft^2}$$

From Eqn. (3.11),

$$V^2 = \frac{7.031 \pm \sqrt{7.031^2 - \frac{93.75^2}{16^2}}}{0.001756 \times 0.02} = \frac{7.031 \pm 3.886}{3.512 \times 10^{-5}}\frac{ft^2}{s^2}$$

$$V = 557.5\frac{ft}{s}\ \ or\ \ 299.2\frac{ft}{s}$$

<u>*Example 3.4*</u>

An airplane has a wing loading of $W/S = 100lb/ft^2$, $C_{D0} = 0.02$, and $K = 0.05$. Draw a T/S versus V plot at an altitude of 10000ft ($\rho = 0.001756slug/ft^3$).

For this plot, we will use Eqn. (3.7).

$$\frac{T}{S} = \frac{1}{2}\rho V^2 C_{D0} + K\frac{2\left(W/_S\right)^2}{\rho V^2}$$

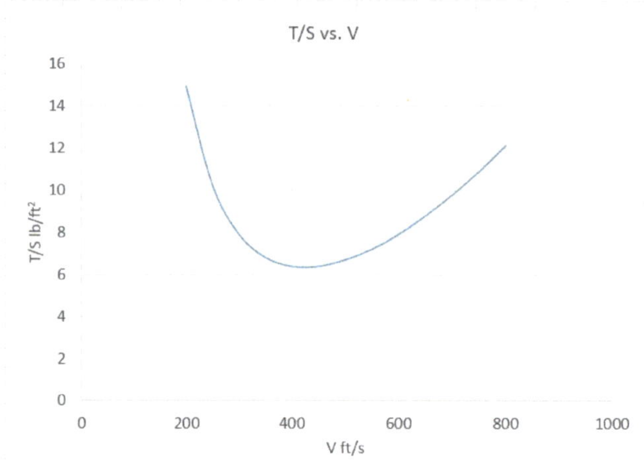

Figure 3.2: *T/S* versus *V*.

Example 3.5
An airplane has a wing loading of W/S = 100lb/ft², C_{D0} = 0.02, and K = 0.05. Determine the minimum thrust and the corresponding airspeed at an altitude of 10000ft (ρ = 0.001756slug/ft³).

From Eqn. (3.10),

$$E_M = \sqrt{\frac{1}{4KC_{D0}}} = \sqrt{\frac{1}{4 \times 0.05 \times 0.02}} = 15.81$$

From Eqn. (3.13),

$$\frac{T}{S} = \frac{W/S}{E_M} = \frac{100}{15.81} = 6.325\frac{lb}{ft^2}$$

From Eqn. (3.11),

$$V = \sqrt{\frac{T/S}{\rho C_{D0}}} = \sqrt{\frac{6.325}{0.001756 \times 0.02}} = 424.4\frac{ft}{s}$$

3.3 Stable and Unstable Branches of the Thrust Curve

We can divide the U-shaped thrust curve into two parts – the left branch and the right branch. We will show that the left branch is statically unstable and the right branch is statically stable.

3.3.1 Unstable, Left Branch

Consider an airplane in level flight with constant airspeed V_a and the corresponding thrust T_a. A gusty headwind V_{head} changes the effective airspeed to V_b.

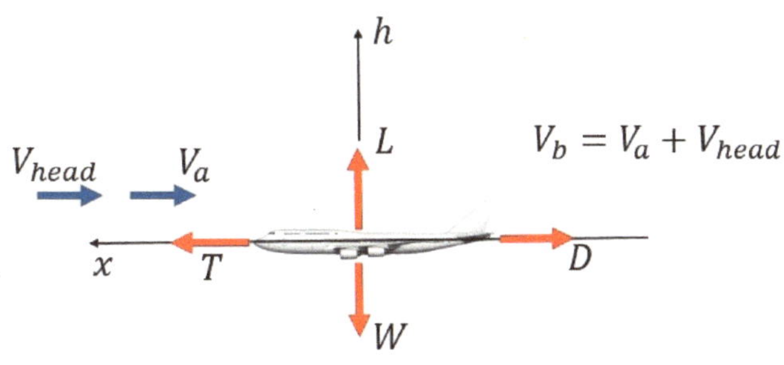

Figure 3.3: Headwind and left branch

The thrust required to maintain the equilibrium of the level flight is $T_b < T_a$ and the pilot has to reduce thrust to T_b. Without any pilot action, the thrust setting T_a is more than the required thrust T_b. The extra thrust further increases the airspeed and the flight state moves farther away from the equilibrium (V_a, T_a).

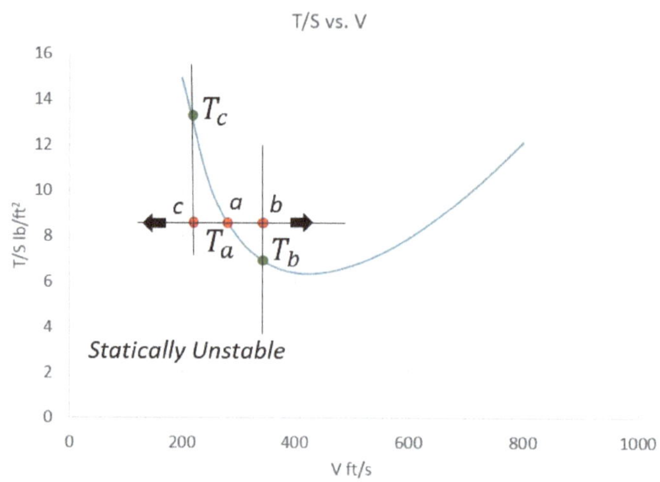

Figure 3.4: Instability of left branch

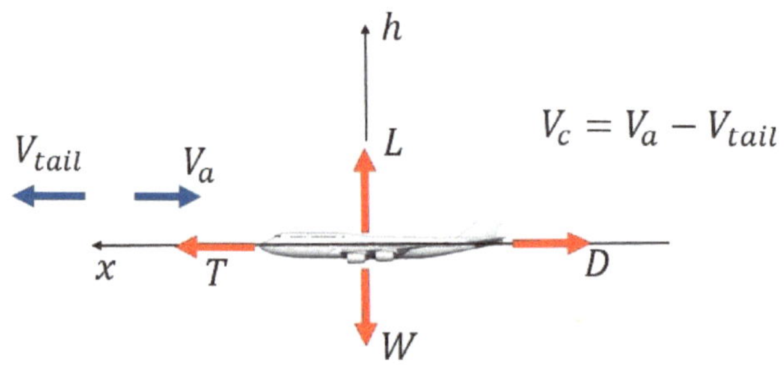

Figure 3.5: Tailwind and left branch

When a gusty tailwind V_{tail} changes the airspeed to V_c, the thrust required to maintain the equilibrium of the level flight $T_c > T_a$. Without any pilot action, the thrust setting T_a is less than the required thrust T_c. The deficient thrust further decreases the airspeed and the flight state moves farther away from the equilibrium (V_a, T_a).

3.3.2 Stable, Right Branch

Consider an airplane in level flight with constant airspeed V_d and the corresponding thrust T_d. A gusty headwind V_{head} changes the effective airspeed to V_e.

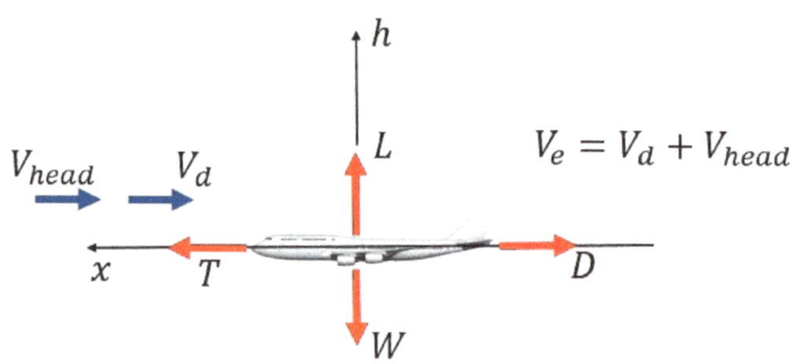

Figure 3.6: Headwind and right branch

The thrust required to maintain the equilibrium of the level flight is $T_e > T_d$. Without any pilot action, the thrust setting T_d is less than the required thrust T_e. The deficient thrust decreases the airspeed and the flight state moves closer to the equilibrium (V_d, T_d).

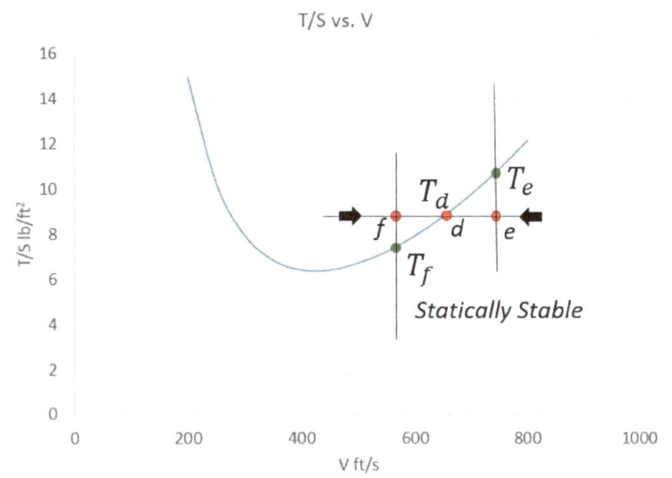

Figure 3.7: Stability of right branch

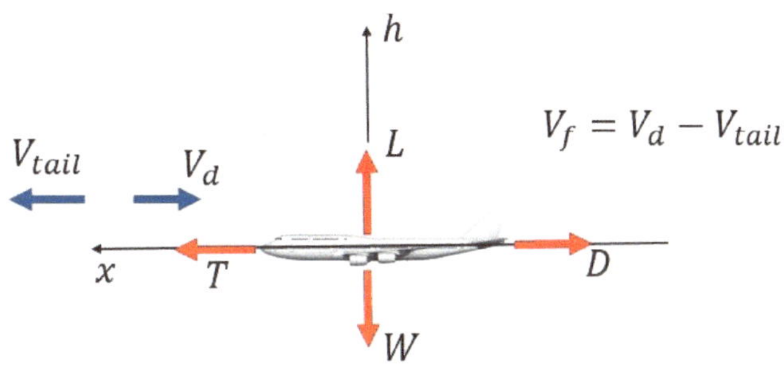

Figure 3.8: Tailwind and right branch

When a gusty tailwind V_{tail} changes the airspeed to V_f, the thrust required to maintain the equilibrium of the level flight $T_f < T_d$. Without any pilot action, the thrust setting T_d is more than the required thrust T_f. The extra thrust increases the airspeed and the flight state moves closer to the equilibrium (V_d, T_d).

3.3 Thrust from Turbojets

The state of flight depends on the characteristics of the airplane, the characteristics of the turbojet engine, and the atmospheric conditions at the flight altitude. We can fix the airplane characteristics by defining the wing geometry. We can fix the characteristics of the engine by defining the inlet area, the pressure ratio across the compressor, and the turbine inlet temperature. We can fix the atmospheric conditions by specifying the flight altitude. In this situation, we determine the state of flight by solving the equation

$$T(V) = D(V)$$

The function *T(V)* is the thrust from a turbojet for the specific engine that we chose for our airplane. The function *D(V)* is the drag for the specific wing that we chose for our airplane. The solution of the equation is a thrust airspeed pair for a specific altitude. The solution is more complex than that given in Eqn. (3.11).

We will simply state the nature of the "flight state" problem, but not delve too deep into this topic. In Section-8.6, we discuss the evaluation of the function *T(V)*. For *D(V)*, we use Eqn. (3.7). The two intersections of the *T(V)* and *D(V)* plots are the two possible states of flight for a specific airplane, a specific engine, and at a specific altitude.

We have drawn the two plots for the following conditions:

- Altitude: *10000ft*
- Airplane: $W = 160000 lb$, $S = 1600 ft^2$, $K = 0.05$, $C_{D0} = 0.02$
- Engine: *Inlet Area* $= 10ft^2$, *Pressure Ratio* $= 12$, *Turbine Inlet Temperature* $= 1600^oR$

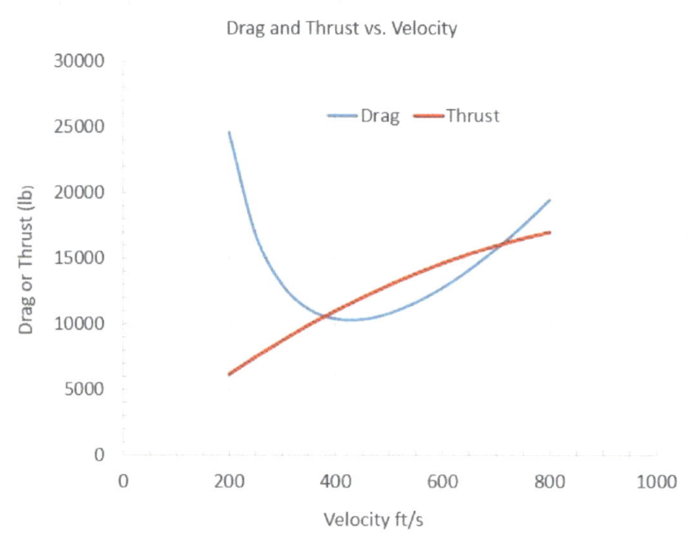

3.3 Absolute Ceiling

For a fixed operating conditions, the thrust from a turbojet engine decreases with the increase of altitude (see Example 8.2 and Eqn. (8.19)). The ratio of thrust at an altitude and thrust at sea level is equal to the ratio of density at that altitude and the density at sea level.

$$\frac{T}{T_{SL}} = \frac{\rho}{\rho_{SL}} \tag{3.15}$$

As the flight altitude of an airplane increases, the thrust decreases and reaches the state of level flight with minimum thrust (see Section 3.2.2). Above this altitude, level flight in equilibrium is impossible (see Section 3.2.1). By combining Eqns. (3.13) and (3.15),

$$\frac{T}{W} = \frac{\rho T_{SL}}{\rho_{SL} W} = \frac{1}{E_M} \; ; \quad \frac{\rho}{\rho_{SL}} = \frac{W}{E_M T_{SL}} \tag{3.16}$$

We can determine the absolute ceiling of the airplane from the density ratio.

<u>Example 3.6</u>
A 300000lb airplane has $E_M = 16$. The thrust from the engines at sea level is $T_{SL} =$ 50000lb. Determine the absolute ceiling. ($\rho_{SL} = 0.002377$ slug/ft^3)

From Eqn. (3.16),

$$\rho = \frac{0.002377 \times 300000}{16 \times 50000} = 0.0008914 \frac{slug}{ft^3}$$

From the standard atmosphere table, this density corresponds to the altitude of 30000ft.

<u>Example 3.7</u>
For an 120000lb airplane with $E_M = 16$, we want the absolute ceiling to be 15000ft. Determine the engine thrust at sea level.
($\rho_{SL} = 0.002377$slug/ft^3, $\rho_{15000} = 0.001496$ slug/ft^3)

From Eqn. (3.16),

$$T_{SL} = \frac{0.002377 \times 120000}{0.001496 \times 16} = 11920lb$$

3.4 Range

An airplane burns fuel during cruise and its weight decreases. The amount of fuel burnt in unit time is proportional to the thrust.

$$-\frac{dW}{dt} \propto T \ ; \ -\frac{dW}{dt} = cT \tag{3.17}$$

The negative sign signifies that the weight of the airplane is decreasing with time. The constant of proportionality c is *TSFC,* the thrust specific fuel consumption (see Example-8.3). *TSFC* is usually given in the unit of *lb/h.lb = 1/h.* By dividing Eqn. (3.3) by Eqn. (3.17) and by using Eqn. (3.2), we get

$$\frac{dx}{dW} = -\frac{V}{cT} = -\frac{V}{cD} \ ; \ dx = -\frac{V}{cD} dW \tag{3.18}$$

Let W_1 and W_2 be the weights of the airplane at the beginning and the end of the cruise, respectively. The amount of fuel burnt is

$$\Delta W_f = W_1 - W_2$$

We can determine the volume of fuel by using the specific weight of jet fuel, *6.8lbs per US gallon.* We define the "cruise fuel weight fraction" as

$$\zeta = \frac{\Delta W_f}{W_1} = 1 - \frac{W_2}{W_1} \tag{3.19}$$

From Eqn. (3.6),

$$W = \frac{1}{2}\rho V^2 S C_L \tag{3.20}$$

As weight decreases, the right-hand-side of Eqn. (3.20) must also decrease. The pilot can achieve this in three possible ways:

1. Keep V and C_L constant and reduce ρ by allowing the airplane to rise to higher altitude.
2. Keep C_L and ρ constant and reduce V by reducing the speed of the airplane.
3. Keep V and ρ constant and reduce C_L by reducing the angle of attack of the wing.

3.4.1 V and C_L Constant Flight

We rewrite Eqn. (3.18) as

$$dx = -\frac{1}{c}\frac{V}{L}\frac{L}{D}dW = -\frac{1}{c}\frac{V}{W}\frac{C_L}{C_D}dW = -\frac{1}{c}VE\frac{dW}{W} \tag{3.21}$$

The ratio E of the drag and lift coefficients is a constant. By integrating Eqn. (3.21), we can find the range X as

$$X_{V,C_L} = -\frac{1}{c}VE\int_1^2 \frac{dW}{W} = -\frac{1}{c}VE\,\ell n\left(\frac{W_2}{W_1}\right)$$

By using Eqn. (3.19),

$$X_{V,C_L} = -\frac{1}{c}VE\,\ell n(1-\zeta) \tag{3.22}$$

In the right-hand-side of Eqn. (3.22), only two quantities have units: V is *distance/time* and c is *1/time.* Hence, if we insert V in *miles/h* and c in *1/h*, we can obtain X in *miles.*

Example 3.8

An airplane has $C_L = 0.42$, $C_{D0} = 0.02$, induced drag coefficient $K = 0.05$, thrust specific fuel consumption $c = 0.6lb/h.lb$, cruise fuel weight fraction $\zeta = 0.3$. Determine the range at a speed of 450miles/h.

$$C_D = C_{D0} + KC_L^2 = 0.02882 \;;\; E = \frac{C_L}{C_D} = 14.57$$

$$X_{V,C_L} = -\frac{450 \times 14.57}{0.6}\,\ell n(1-0.3) = 3898\,miles$$

Example 3.9

A 250000lb airplane has $C_L = 0.42$, $C_{D0} = 0.02$, induced drag coefficient $K = 0.05$, thrust specific fuel consumption $c = 0.6lb/h.lb$. Determine the weight of fuel required for a range of 2000miles at a speed of 450miles/h.

$$C_D = C_{D0} + KC_L^2 = 0.02882 \ ; \ E = \frac{C_L}{C_D} = 14.57$$

$$2000 = -\frac{450 \times 14.57}{0.6} \ell n(1-\zeta) \ ; \ \ell n(1-\zeta) = -0.1830 \ ; \ \zeta = 0.1672$$

From Eqn. (3.19),

$$\Delta W_f = 0.1672 \times 250000 = 41810 \ lb$$

Example 3.10

A 250000lb airplane has wing area $S = 2600ft^2$, $C_{D0} = 0.02$, induced drag coefficient $K = 0.05$, thrust specific fuel consumption $c = 0.6lb/h.lb$. The airplane starts cruising at an altitude of 30000ft with a speed of 500miles/h. Determine the altitude after the airplane cruises for 3000miles. (ρ at 30000ft =0.0008907slug/ft^3)

$$V = \frac{500 \times 5280}{3600} = 733.3\frac{ft}{s}$$

$$C_L = \frac{250000}{0.5 \times 0.0008907 \times 733.3^2 \times 2600} = 0.4015$$

$$C_D = C_{D0} + KC_L^2 = 0.02806 \ ; \ E = \frac{C_L}{C_D} = 14.31$$

$$3000 = -\frac{733.3 \times 14.31}{0.6} \ell n(1-\zeta) ; \ell n(1-\zeta) = -0.1715 ; \zeta = 0.1576$$

$$\Delta W_f = 0.1576 \times 250000 = 39400 \ lb$$

$$W_2 = W_1 - \Delta W_f = 210600 \ lb$$

$$\rho_2 = \frac{210600}{0.5 \times 733.3^2 \times 0.4015 \times 2600} = 0.0007504\frac{slug}{ft^3}$$

Altitude corresponding to ρ_2 is about 34500ft. From the beginning of the cruise to the end, the airplane climbs about 4500ft.

3.4.2 ρ and C_L Constant Flight

By differentiating Eqn. (3.20),

$$dW = \varrho V S C_L dV \tag{3.23}$$

By inserting Eqns. (3.20) and (3.23) in Eqn. (3.21)

$$dx = -\frac{1}{c} VE \frac{\varrho V S C_L dV}{\frac{1}{2}\varrho V^2 S C_L} = -\frac{2E}{c} dV$$

By integrating,

$$X_{\varrho,C_L} = -\frac{2E}{c}(V_2 - V_1) = \frac{2EV_1}{c}\left(1 - \frac{V_2}{V_1}\right) \tag{3.24}$$

From Eqn. (3.20), at the beginning and the end of the cruise

$$\frac{W_2}{W_1} = \frac{\frac{1}{2}\varrho V_2^2 S C_L}{\frac{1}{2}\varrho V_1^2 S C_L} = \frac{V_2^2}{V_1^2} \; ; \; \frac{V_2}{V_1} = \left(\frac{W_2}{W_1}\right)^{1/2} \tag{3.25}$$

By combining Eqns. (3.19) and (3.25),

$$\frac{V_2}{V_1} = (1 - \zeta)^{1/2} \tag{3.26}$$

By inserting Eqn. (3.26) in Eqn. (3.24),

$$X_{\varrho,C_L} = \frac{2EV_1}{c}\left\{1 - (1 - \zeta)^{1/2}\right\} \tag{3.27}$$

In the right-hand-side of Eqn. (3.27), only two quantities have units: V_1 is *distance/time* and c is *1/time*. Hence, if we insert V_1 in *miles/h* and c in *1/h*, we can obtain X in *miles.*

Example 3.11

An airplane has $C_L = 0.42$, $C_{D0} = 0.02$, induced drag coefficient $K = 0.05$, thrust specific fuel consumption $c = 0.6lb/h.lb$, cruise fuel weight fraction $\zeta = 0.3$. Determine the range when the speed is 450 miles/h at the beginning of the cruise. Also, determine the speed at the end of the cruise.

$$C_D = C_{D0} + KC_L^2 = 0.02882 \; ; \; E = \frac{C_L}{C_D} = 14.57$$

$$X_{\rho,C_L} = \frac{2 \times 450 \times 14.57}{0.6}\left\{1 - (1 - 0.3)^{1/2}\right\} = 3570 \; miles$$

From Eqn. (3.26),

$$V_2 = 450 \times (1 - 0.3)^{1/2} = 376.5\frac{miles}{h}$$

3.4.2 V and ρ Constant Flight

We write Eqn. (3.18) as

$$dx = -\frac{V}{c}\frac{dW}{QSC_D} = -\frac{V}{cQS}\frac{dW}{(C_{DO} + KC_L^2)} \tag{3.28}$$

The dynamic pressure is Q. From Eqn. (3.20),

$$W = \frac{1}{2}\rho V^2 SC_L = QSC_L \;;\; C_L = \frac{W}{QS} \tag{3.29}$$

By inserting Eqn. (3.29) in Eqn. (3.28),

$$dx = -\frac{V}{cQS}\frac{dW}{C_{DO} + K\left(\frac{W^2}{Q^2S^2}\right)} = -\frac{VQS}{cK}\frac{dW}{\left\{\frac{Q^2S^2C_{DO}}{K} + W^2\right\}}$$

By using the standard integration formula,

$$\int\frac{dz}{a^2 + z^2} = \frac{1}{a}tan^{-1}\left(\frac{z}{a}\right) + Const.$$

$$X_{V,\rho} = \frac{V}{c}\frac{1}{\sqrt{KC_{DO}}}\left\{tan^{-1}\left(\frac{W_1}{QS}\sqrt{\frac{K}{C_{DO}}}\right) - tan^{-1}\left(\frac{W_2}{QS}\sqrt{\frac{K}{C_{DO}}}\right)\right\}$$

By using the trigonometric identity

$$tan^{-1}\alpha - tan^{-1}\beta = tan^{-1}\frac{\alpha - \beta}{1 + \alpha\beta}$$

$$X_{V,\rho} = \frac{V}{c}\frac{1}{\sqrt{KC_{DO}}}\,tan^{-1}\left\{\frac{\left(\frac{W_1}{QS}\sqrt{\frac{K}{C_{DO}}}\right) - \left(\frac{W_2}{QS}\sqrt{\frac{K}{C_{DO}}}\right)}{1 + \frac{W_1W_2}{Q^2S^2}\frac{K}{C_{DO}}}\right\} \tag{3.30}$$

We use Eqn. (3.19) to simplify the numerator of Eqn. (3.30) as

$$\left(\frac{1}{QS}\right)\left(\sqrt{\frac{K}{C_{D0}}}\right)(W_1 - W_2) = \left(\frac{1}{QS}\right)\left(\sqrt{\frac{K}{C_{D0}}}\right)\zeta W_1 = \left(\sqrt{\frac{K}{C_{D0}}}\right)\zeta C_{L1}$$

C_{L1} is the lift coefficient at the beginning of the cruise. We use Eqn. (3.19) to simplify the denominator of Eqn. (3.30) as

$$1 + \frac{W_1^2(1-\zeta)}{Q^2S^2}\frac{K}{C_{D0}} = 1 + (1-\zeta)\left(\frac{K}{C_{D0}}\right)C_{L1}^2$$

Then

$$X_{V,\rho} = \frac{V}{c}\frac{1}{\sqrt{KC_{D0}}}\tan^{-1}\left\{\frac{\left(\sqrt{\frac{K}{C_{D0}}}\right)\zeta C_{L1}}{1 + (1-\zeta)\left(\frac{K}{C_{D0}}\right)C_{L1}^2}\right\}$$

Or

$$X_{V,\rho} = \frac{V}{c}\frac{1}{\sqrt{KC_{D0}}}\tan^{-1}\left\{\frac{\left(\sqrt{KC_{D0}}\right)\zeta C_{L1}}{C_{D0} + (1-\zeta)KC_{L1}^2}\right\} \tag{3.30}$$

By using the relation,

$$C_{L1} = \frac{2W_1}{\rho V^2 S} \; ; \; C_{D1} = C_{D0} + KC_{L1}^2 \tag{3.31}$$

C_{D1} is the drag coefficient at the beginning of the cruise. Then

$$X_{V,\rho} = \frac{V}{c\sqrt{KC_{D0}}}\tan^{-1}\left\{\frac{\left(\sqrt{KC_{D0}}\right)\zeta C_{L1}}{C_{D1} - \zeta KC_{L1}^2}\right\} \tag{3.32}$$

In the right-hand-side of Eqn. (3.32), only two quantities have units: V is *distance/time* and c is *1/time.* Hence, if we insert V in *miles/h* and c in *1/h*, we can obtain X in *miles.*

<u>Example 3.12</u>

A 250000lb airplane has wing area $S = 2400ft^2$, $C_{D0} = 0.02$, induced drag coefficient $K = 0.05$, thrust specific fuel consumption $c = 0.6lb/h.lb$, cruise fuel weight fraction $\zeta = 0.3$. Determine the range at a speed of 450miles/h at an altitude of 20000ft. Also, determine the lift coefficient at the end of the cruise.
(ρ at 20000ft =0.001267slug/ft^3).

$$V = \frac{450 \times 5280}{3600} = 660\frac{ft}{s}$$

$$C_{L1} = \frac{250000}{0.5 \times 0.001267 \times 660^2 \times 2400} = 0.3775$$

$$W_2 = 250000 \times (1 - 0.3) = 175000\ lb$$

$$C_{L2} = \frac{175000}{0.5 \times 0.001267 \times 660^2 \times 2400} = 0.2643$$

$$C_{D1} = 0.02 + 0.05 \times 0.3775^2 = 0.02712$$

$$X_{V,\rho} = \frac{450}{0.6} \times \frac{1}{\sqrt{0.05 \times 0.02}} \times tan^{-1}\left\{\frac{\sqrt{0.05 \times 0.02} \times 0.3 \times 0.3775}{0.02712 - 0.3 \times 0.05 \times 0.3775^2}\right\}$$

$$X_{V,\rho} = 3375\ miles$$

Flights under the regulation of the air traffic control use the V and ρ constant flight program. Due to the complexity of Eqn. (3.32), determination of the fuel consumption ζ requires lengthy hand calculation. However, we can easily do the calculations for this flight program by using Excel spreadsheet (see Appendix C).

3.5 Best Range

To find the best range, we need to find V_{BR} the airspeed of the airplane for best range. We do this by maximizing the integrand of Eqn. (3.18). At the maximum

$$\frac{d}{dV}\left(\frac{V}{cD}\right) = 0$$

Or

$$-\frac{V}{cD^2}\frac{dD}{dV} + \frac{1}{cD} = 0\ ;\ \frac{dD}{dV} = \frac{D}{V} \tag{3.33}$$

By using Eqn. (3.7) in Eqn. (3.33),

$$\rho V_{BR} S C_{DO} - \frac{4KS\left(W/_S\right)^2}{\rho V_{BR}^3} = \frac{1}{2}\rho V_{BR} S C_{DO} + \frac{2KS\left(W/_S\right)^2}{\rho V_{BR}^3}$$

Or

$$\frac{6KS\left(W/_S\right)^2}{\rho V_{BR}^3} = \frac{1}{2}\rho V_{BR} S C_{DO}\ ;\ V_{BR} = \left(\frac{12K\left(W/_S\right)^2}{\rho^2 C_{DO}}\right)^{1/4} \tag{3.34}$$

Observation:

- The V_{BR} of Eqn. (3.34) is the instantaneous best range airspeed corresponding to the weight of the airplane at that instant. At the next instant, when the weight of the airplane changes due to fuel consumption, V_{BR} also changes.

From Eqn. (3.6) and Eqn. (3.34), the lift coefficient corresponding to V_{BR} is

$$C_{L,BR} = \frac{W}{\frac{1}{2}\rho V_{BR}^2 S} = \frac{\left(W/_S\right)}{\frac{1}{2}\rho \frac{\left(W/_S\right)}{\rho}\left(\frac{12K}{C_{D0}}\right)^{\frac{1}{2}}} = \left(\frac{C_{D0}}{3K}\right)^{\frac{1}{2}} \qquad (3.35)$$

Observation:

- To maximize the range at all instants of the airplane's journey from origin to destination, we need to keep the lift coefficient constant at all times.

The corresponding drag coefficient is

$$C_{D,BR} = C_{D0} + KC_{L,BR}^2 = \frac{4}{3}C_{D0} \qquad (3.36)$$

The lift to drag ratio is

$$E_{BR} = \frac{C_{L,BR}}{C_{D,BR}} = \left(\frac{C_{D0}}{3K}\right)^{\frac{1}{2}}\left(\frac{3}{4C_{D0}}\right) = \left(\frac{3}{16KC_{D0}}\right)^{\frac{1}{2}} \qquad (3.37)$$

As lift coefficient remains constant during cruise with best range, strictly speaking, we can use Eqn. (3.37) only for the flight programs in Sections 3.4.1 and 3.4.2. From Eqn. (3.22),

$$X_{BR,V,C_L} = -\frac{V_{BR}E_{BR}}{c}\,\ell n(1-\zeta) \qquad (3.38)$$

From Eqn. (3.27)

$$X_{BR,\varrho,C_L} = \frac{2E_{BR}V_{BR,1}}{c}\left\{1-(1-\zeta)^{1/2}\right\} \qquad (3.39)$$

<u>*Example 3.13*</u>

An airplane has wing loading $W/S = 90\ lb/ft^2$ at the beginning of a cruise, $C_{D0} = 0.02$, induced drag coefficient $K = 0.05$, thrust specific fuel consumption $c = 0.6 lb/h.lb$, cruise fuel weight fraction $\zeta = 0.3$. Determine the best ranges $X_{BR,V,CL}$ and $X_{BR,\rho,CL}$ at an altitude of 20000ft.

(ρ at 20000ft =0.001267slug/ft^3)

From Eqn. (3.34),

$$V_{BR} = \left(\frac{12 \times 0.05 \times 90^2}{0.001267^2 \times 0.02}\right)^{1/4} = 623.8\frac{ft}{s} = 425.3\frac{miles}{h}$$

$$E_{BR} = \left(\frac{3}{16 \times 0.05 \times 0.02}\right)^{\frac{1}{2}} = 13.69$$

$$X_{BR,V,C_L} = -\frac{425.3 \times 13.69}{0.6}\,\ell n(1-0.3) = 3461\,miles$$

$$X_{BR,\varrho,C_L} = \frac{2 \times 13.69 \times 425.3}{0.6}\left\{1-(1-0.3)^{1/2}\right\} = 3170\,miles$$

3.6 Endurance

Endurance is the longest time an airplane can cruise for a given fuel consumption and given airplane characteristics. We rewrite Eqn. (3.17) as

$$\frac{dt}{dW} = -\frac{1}{cT} = -\frac{1}{cD} = -\frac{1}{c}\frac{L}{D}\frac{1}{L} = -\frac{1}{c}\frac{E}{L} = -\frac{1}{c}\frac{E}{W}$$

Or

$$dt = -\frac{E}{c}\frac{dW}{W} \tag{3.40}$$

In Eqn. (3.40), we maximize the duration of level flight by maximizing E. When we maximize E to E_M, the endurance is

$$t_{max} = -\frac{E_M}{c}\int_1^2 \frac{dW}{W} = -\frac{E_M}{c}\,\ell n\left(\frac{W_2}{W_1}\right) = \frac{E_M}{c}\,\ell n\left(\frac{1}{1-\zeta}\right) \tag{3.41}$$

The unit of c is *1/h* and unit of t_{max} from Eqn. (3.40) is h. By using Eqn. (2.40), the airspeed corresponding to the maximum flight time is

$$V_{tmax} = \sqrt{\frac{2(W/S)}{\varrho C_{L,EM}}} = \left\{\frac{2(W/S)}{\rho}\right\}^{1/2}\left(\frac{K}{C_{D0}}\right)^{1/4} \tag{3.42}$$

Example 3.14

Determine the endurance of an airplane with C_{D0} = 0.02, induced drag coefficient K = 0.05, thrust specific fuel consumption c = 0.6lb/h.lb, and cruise fuel weight fraction $\zeta =$ 0.5.

From Eqn. (2.41),

$$E_M = \frac{1}{\sqrt{4 \times 0.05 \times 0.02}} = 15.81$$

$$t_{max} = \frac{15.81}{0.6}\, ln\left(\frac{1}{1-0.5}\right) = 18.26\ hours$$

4 Takeoff

4.1 Introduction

The takeoff operation consists of two parts: ground run and climb. Large long-haul airplanes need longer ground run compared to the small short-hop airplanes. In addition, length of ground run increases with the altitude of the airport. Length and altitude of the runway at the airports, that we want our airplane to serve, affect our design decisions, such as, the engine size, the wing loading, and the high lift devices in the airplane. Furthermore, our design decisions must also include unpredictable situations such as the condition of the runway, engine failure, and aborted takeoff.

For the climb, we will consider the fastest climb as a performance characteristic.

4.2 Ground Run

4.2.1 Equations of Motion

Forces acting on an airplane, during ground run, are thrust T, drag D, lift L, takeoff weight W_{TO}, normal force from the runway N, and force from rolling friction F. Velocity and acceleration of the airplane are V and a, respectively.

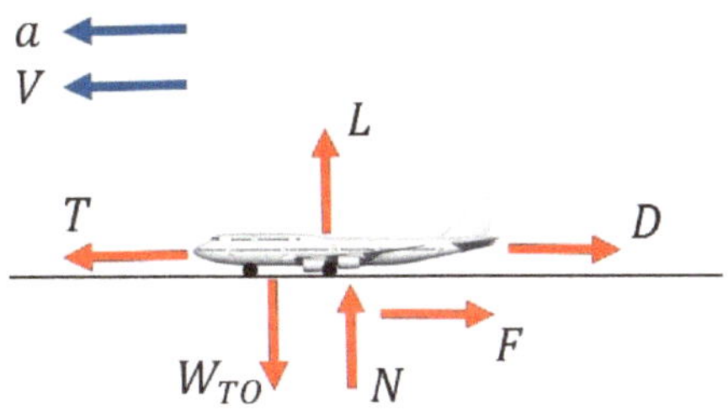

Figure 4.1: Ground run free-body-diagram

The vertical force equilibrium and the friction force are

$$N = W_{TO} - L \ ; \quad F = \mu N = \mu(W_{TO} - L)$$

The friction coefficient is μ. The horizontal equation of motion is

$$T - D - \mu(W_{TO} - L) = \frac{W_{TO}}{g} a = \frac{W_{TO}}{g}\frac{dV}{dt}$$

By writing lift and drag in terms of the lift and drag coefficients, the equation of motion becomes

$$\frac{g}{W_{TO}}(T - \mu W_{TO}) + \frac{1}{2}\frac{g}{W_{TO}}\rho V^2 S(\mu C_L - C_D) = \frac{dV}{dt} \tag{4.1}$$

We define

$$A = \frac{g}{W_{TO}}(T - \mu W_{TO}) \; ; \; B = \frac{1}{2}\frac{g}{W_{TO}}\rho S(\mu C_L - C_D) \tag{4.2}$$

By combining Eqns. (4.1) and (4.2),

$$A + BV^2 = \frac{dV}{dt} = \frac{dV}{dx}\frac{dx}{dt} = V\frac{dV}{dx} \tag{4.3}$$

By rearranging Eqn. (4.3),

$$dx = \frac{VdV}{A + BV^2} \tag{4.4}$$

4.2.2 Ground Run and Takeoff Velocity

To integrate Eqn. (4.4), we use the substitution

$$z = A + BV^2 \; ; \; VdV = \frac{dz}{2B} \tag{4.5}$$

With this substitution, Eqn. (4.4) becomes

$$dx = \frac{1}{2B}\frac{dz}{z} \tag{4.6}$$

By integrating Eqn. (4.6), we find the ground run as

$$d = \int_{start}^{TO} dx = \frac{1}{2B}\int_{start}^{TO}\frac{dz}{z} = \frac{1}{2B}\ell n\left(\frac{z_{TO}}{z_{start}}\right) \tag{4.7}$$

At start, $V = 0$, and at takeoff, $V = V_{TO}$. Thus

$$d = \frac{1}{2B}\ell n\left(\frac{A + BV_{TO}^2}{A}\right) = \frac{1}{2B}\ell n\left(1 + \frac{BV_{TO}^2}{A}\right) \tag{4.8}$$

By using the McLaurin Series,

$$\ell n(1 + u) = u - \frac{u^2}{2} + \frac{u^3}{3} - \cdots$$

The first approximation for the ground run is

$$d = \frac{1}{2B}\frac{BV_{TO}^2}{A} = \frac{V_{TO}^2}{2A} = \frac{W_{TO}}{2g}\frac{V_{TO}^2}{(T - \mu W_{TO})} = \frac{V_{TO}^2}{2g\left\{\left({}^{T}/_{W_{TO}}\right) - \mu\right\}} \tag{4.9}$$

In the figure below, we show the experimental data for the friction coefficient (Wetmore, 1937).

TABLE I.—ROLLING-FRICTION COEFFICIENTS

Wheels	Bearings	Load per wheel	Inflation pressure	Static tire deflection	Rolling-friction coefficient, μ		
					Concrete	Firm turf	Soft turf
		Pounds	*lb. sq. in.*	*Inches*			
Extra low pressure:							
22X10-4	Plain	1,240	12.5	2.55	0.029	0.035	------
		940	12.5	2.06	.030	.041	
		640	12.5	1.58	.025	.054	
		940	10	2.50	.028	.047	
		940	8	2.92	.033	.050	
30X13-6	do	1,740	12.5	2.82	.027	.046	------
		1,340	12.5	2.29	.024	.046	
		940	12.5	1.74	.023	.047	
		940	10	1.90	.029	.049	
		940	8	2.19	.035	.047	
Low pressure:							
7.50-10	Roller	1,540	20	2.26	.013	.025	------
		1,240	20	1.90	.010	.023	
		940	20	1.52	.009	.029	
		940	16	1.78	.010	.026	
		940	12	2.18	.012	.026	
8.50-10	do	1,740	20	2.52	.013	.030	------
		1,340	20	1.93	.014	.031	
		940	20	1.56	.010	.034	
		940	16	1.83	.013	.029	
		940	12	2.22	.015	.030	
	Plain	1,740	20	2.52	.020	-----	------
		940	20	1.56	.018		
Standard:							
26X5	do	1,240	50	.94	.018	-----	0.070
		940	50	.76	.015		.071
		640	50	.58	.013		.066
36X8	do	1,740	60	.80	.017	-----	.064
		1,340	60	.67	.011		.072
		940	60	.53	.015		.077
		940	60	.53	------	.037	------
		940	50	.62	.020	.033	
		940	40	.69	.025	.033	

Figure 4.2: Rolling friction coefficient.

Typically, the takeoff airspeed is about 20% more than the stall airspeed. We determine the stall airspeed at takeoff with high lift devices (see Example 2.19).

$$V_{stall} = \left\{ \frac{2\left({W_{TO}}/{S} \right)}{\rho C_{L,max,TO}} \right\}^{\frac{1}{2}}$$

and

$$V_{TO} = 1.2V_{stall} = \left\{\frac{2.88\left({W_{TO}}/{S}\right)}{\rho C_{L,max,TO}}\right\}^{\frac{1}{2}} \tag{4.10}$$

By combining Eqns. (4.9) and (4.10), we get the ground run as

$$d = \frac{1.44\left({W_{TO}}/{S}\right)}{\rho g C_{L,max,TO}\left\{\left({T}/{W_{TO}}\right) - \mu\right\}} \tag{4.11}$$

Test calculations:

- We should ensure that among all the airports that the airplane serves, the airport at the highest altitude has sufficiently long runway.
- We should calculate the ground run with the highest value of the friction coefficient from Figure 4.2.

Design decisions to reduce the ground run:

- We can reduce the wing loading to reduce the ground run. However, reducing the wing loading reduces the best range airspeed and consequently the best range (see Eqn. 3.33).
- We can increase the thrust to weight ratio by using a bigger engine. However, this increases the overall weight of the airplane and increases the fuel consumption for a given range.
- We can increase the maximum lift coefficient at takeoff. This increases the induced drag and the airplane will require more thrust during the ground run.

Aircraft	TO Weight lb	Max. Thrust (SL) lb	T/W	Area ft.sq.	W/S lb/ft.sq.
Airbus 300	378535	123000	0.325	2800	135.2
Airbus 380	1190500	268000	0.225	9100	130.8
Beechjet 400	16100	5930	0.368	241	66.8
Boeing 737	115500	28000	0.242	1346	85.8
Boeing 747	833000	210000	0.252	5650	147.4
Bombardier CRJ	51000	18440	0.362	587	86.9
Cessna 750	35700	12884	0.361	527	67.7
Gulfstream IV	74600	27700	0.371	950	78.5

Figure 4.3: Thrust to weight ratio and wing loading

Unforeseen situations:

- The ground run from Eqn. (4.11) does not account for engine failure or aborted takeoff. Typically, the ground run is less than the half of the length of the runway.

In Figure 4.3, we have calculated the thrust to weight ratio by using the maximum thrust at the sea level. We cannot calculate the ground run by inserting this ratio in Eqn. (4.11). We have to consider two factors to calculate thrust to weight ratio.

i. The throttle is not set for maximum thrust.
ii. The engine is not operating at the sea level. We must correct the thrust for altitude by using Eqn. (8.19).

<u>*Example 4.1*</u>
Determine the ground run for Airbus 380 with full thrust at sea level and at an altitude of 5000ft. The coefficient of rolling friction is 0.03 and maximum lift coefficient at takeoff is 1.8.

The thrust to weight ratio and the density at sea level are (see Airbus 380 data in Figure 4.3)

$$\frac{T}{W_{TO}} = \frac{T_{SL}}{W_{TO}} = 0.225 \;;\; \rho = 0.002377\frac{slug}{ft^3}$$

$$d = \frac{1.44 \times 130.8}{0.002377 \times 32.2 \times 1.8 \times \{0.225 - 0.03\}} = 7011ft$$

Thrust to weight ratio and density at 5000ft are

$$\frac{T}{W_{TO}} = \frac{T_{SL}}{W_{TO}}\frac{\rho}{\rho_{SL}} = 0.225 \times \frac{0.002048}{0.002377} = 0.1939 \;;\; \rho = 0.002048\frac{slug}{ft^3}$$

$$d = \frac{1.44 \times 130.8}{0.002048 \times 32.2 \times 1.8 \times \{0.1939 - 0.03\}} = 9681ft$$

The length of runway at Denver is 16000ft.

<u>*Example 4.2*</u>
Determine the ground run for Cessna 750 for 60% thrust at the sea level. The coefficient of rolling friction is 0.03 and maximum lift coefficient at takeoff is 1.8.

The thrust to weight ratio and the density at sea level are (see Cessna 750 data in Figure 4.3)

$$\frac{T}{W_{TO}} = 0.6 \times \frac{T_{SL}}{W_{TO}} = 0.6 \times 0.361 = 0.2166 \;;\; \rho = 0.002377\frac{slug}{ft^3}$$

$$d = \frac{1.44 \times 67.7}{0.002377 \times 32.2 \times 1.8 \times \{0.2166 - 0.03\}} = 3792ft$$

4.2 Climb

4.2.1 Climb Angle

The free-body-diagram for a climbing flight is

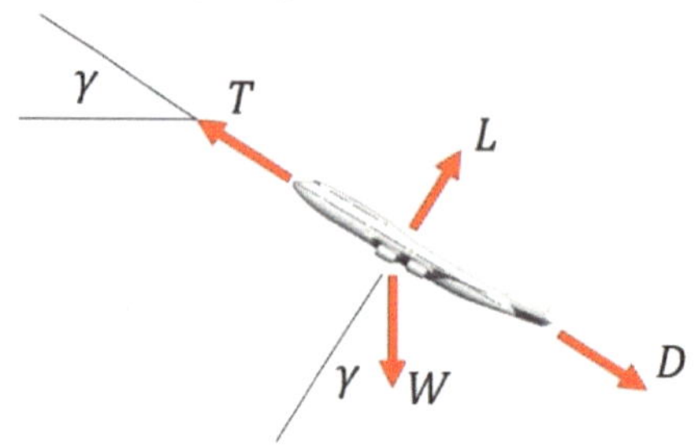

Figure 4.4: Free body diagram for climb

For steady climb, the force balance equations are

$$L = W\cos\gamma \tag{4.12}$$

$$T - D - W\sin\gamma = 0\ ;\quad \sin\gamma = \frac{T}{W} - \frac{D}{W} \tag{4.13}$$

By combining Eqns. (4.12) and (4.13),

$$\sin\gamma = \frac{T}{W} - \frac{D}{W} = \frac{T}{W} - \frac{D}{L}\cos\gamma = \frac{T}{W} - \frac{\cos\gamma}{E} \tag{4.14}$$

The climb angle γ is usually small and we use the approximation

$$\cos\gamma \approx 1$$

By using this approximation, we can determine the climb angle from the equation

$$\sin\gamma = \frac{T}{W} - \frac{1}{E} \tag{4.15}$$

For some problems, an alternative form of Eqn. (4.15) is more useful. By using the definition of dynamic pressure

$$L = QSC_L = W\cos\gamma\ ;\quad C_L = \frac{W\cos\gamma}{QS} \tag{4.16}$$

By using Eqn. (4.16), we can write drag D as

$$D = QS\left(C_{D0} + K\frac{W^2\cos^2\gamma}{Q^2S^2}\right) = QSC_{D0} + K\frac{W^2\cos^2\gamma}{QS} \tag{4.17}$$

By using Eqn. (4.17) in Eqn. (4.13),

$$sin\gamma = \frac{T}{W} - \frac{QC_{DO}}{\left(W/_S\right)} - \frac{K\left(W/_S\right)}{Q}cos^2\gamma \approx \frac{T}{W} - \left\{\frac{QC_{DO}}{\left(W/_S\right)} + \frac{K\left(W/_S\right)}{Q}\right\} \tag{4.18}$$

We must correct the thrust for altitude as

$$\frac{T}{W} = \frac{T_{SL}}{W}\frac{\rho}{\rho_{SL}} \tag{4.19}$$

<u>*Example 4.3*</u>
An airplane has a thrust to weight ratio of 0.25 at sea level, wing loading of 110lb/ft², C_{D0} = 0.03, induced drag coefficient K = 0.06, and airspeed 550ft/s. Determine the climb angle at sea level and at an altitude of 10000ft.

At sea level,

$$\rho = 0.002377\frac{slug}{ft^3} \ ; \ \frac{T}{W} = 0.25$$

Dynamic pressure

$$Q = 0.5 \times 0.002377 \times 550^2 = 359.5\frac{lb}{ft^2}$$

$$sin\gamma = 0.25 - \left\{\frac{359.5 \times 0.03}{110} + \frac{0.06 \times 110}{359.5}\right\} = 0.1336 \ ; \ \gamma = 7.678^o$$

At 10000ft,

$$\rho = 0.001756\frac{slug}{ft^3} \ ; \ \frac{T}{W} = 0.25 \times \frac{0.001756}{0.002377} = 0.1847$$

Dynamic pressure

$$Q = 0.5 \times 0.001756 \times 550^2 = 265.6\frac{lb}{ft^2}$$

$$sin\gamma = 0.1847 - \left\{\frac{265.6 \times 0.03}{110} + \frac{0.06 \times 110}{265.6}\right\} = 0.08741 \ ; \ \gamma = 5.015^o$$

For the same airspeed, the climb angle decreases with altitude.

We can calculate the airspeed for a given climb angle. We can perform this complex calculation by using the Solver Add-In available in Excel Spreadsheet (see Example D.1 in Appendix D).

4.2.2 Rate of Climb

We calculate the rate of climb in *ft/s* from

$$\dot{h} = Vsin\gamma \tag{4.20}$$

Example 4.4

An airplane has a thrust to weight ratio of 0.25 at sea level, wing loading of 110lb/ft², $C_{D0} = 0.03$, induced drag coefficient K = 0.06, and airspeed 550ft/s. Determine the rate of climb at sea level and at an altitude of 10000ft.

From Example 4.3,

$$sin\gamma_{SL} = 0.1336 \ ; \ \dot{h}_{SL} = 550 \times 0.1336 = 73.48\frac{ft}{s} in\gamma_{10000} = 0.08741 \ ; \ \dot{h}_{10000} = 550 \times 0.08741 = 48.08\ ft/s$$

4.2.3 Fastest Climb

For fastest climb, we maximize the rate of climb by calculating the corresponding airspeed. At the maximum rate of climb

$$\frac{d\dot{h}}{dV} = \frac{d}{dV}(Vsin\gamma) = \frac{d}{dV}\left(\frac{TV - DV}{W}\right) = \frac{1}{W}\left(T - D - V\frac{dD}{dV}\right) = 0 \tag{4.21}$$

By using Eqn. (4.17), Eqn. (4.21) becomes

$$T - QSC_{D0} - \frac{KW^2}{QS} - V\frac{dQ}{dV}\left(SC_{D0} - \frac{KW^2}{Q^2S}\right) = 0 \tag{4.22}$$

Also

$$\frac{dQ}{dV} = \frac{d}{dV}\left(\frac{1}{2}\rho V^2\right) = \rho V = \frac{2Q}{V} \tag{4.23}$$

By inserting Eqn. (4.23) in Eqn. (4.22),

$$T - 3Q_{FC}SC_{D0} + \frac{KW^2}{Q_{FC}S} = 0$$

Or

$$3Q_{FC}^2S^2C_{D0} - TQ_{FC}S - KW^2 = 0 \tag{4.24}$$

By solving the quadratic equation (4.24) for Q and retaining on the realistic, positive solution

$$Q_{FC} = \frac{T + \sqrt{T^2 + 12KC_{D0}W^2}}{6SC_{D0}}$$

By using Eqn. (2.41),

$$Q_{FC} = \frac{T}{6SC_{D0}} + \frac{T}{6SC_{D0}}\sqrt{1 + \left(3W^2/T^2E_M^2\right)}$$

Or

$$Q_{FC} = \frac{(T/S)}{6C_{D0}}\left[1 + \sqrt{1 + \frac{3}{\{E_M(T/W)\}^2}}\right] \tag{4.25}$$

The dynamic pressure corresponding to the fastest climb changes with altitude because the thrust to weight ratio changes with altitude (see Eqn. (4.19)). We can calculate the airspeed corresponding to the fastest climb as

$$V_{FC} = \sqrt{\frac{2Q_{FC}}{\rho}} \tag{4.26}$$

From Eqn. (4.17), the drag corresponding to the fastest climb is

$$D_{FC} = Q_{FC}SC_{D0} + K\frac{W^2}{Q_{FC}S} \tag{4.27}$$

We calculate the fastest climb rate from

$$sin\gamma_{FC} = \frac{(T - D_{FC})}{W} \; ; \; \dot{h}_{FC} = V_{FC}sin\gamma_{FC} \tag{4.28}$$

We have shown the derivation for maximizing the rate of climb for the completeness of the text. However, we can do this maximization of rate of climb without any derivation or long calculations by using the Solver Add-In available in Excel spreadsheet (see Example D.2 in Appendix D).

<u>*Example 4.5*</u>
A 200000lb airplane has a wing area of 2000ft², C_{D0} = 0.02, K = 0.05, thrust to weight ratio at sea level of 0.31. Determine fastest rate of climb at sea level and at an altitude of 20000ft.

For sea level,

$$\rho = 0.002377\frac{slug}{ft^3} \; ; \; \frac{T}{W} = 0.31$$

$$E_M = \sqrt{\frac{1}{4KC_{D0}}} = 15.81 \; ; \; \frac{T}{S} = \frac{T}{W}\frac{W}{S} = 0.31 \times 100 = 31.0\frac{lb}{ft^2}$$

$$Q_{FC} = \frac{31}{6 \times 0.02}\left[1 + \sqrt{1 + \frac{3}{\{15.81 \times 0.31\}^2}}\right] = 532.3\frac{lb}{ft^2}$$

$$V_{FC} = \sqrt{\frac{2 \times 532.3}{0.002377}} = 669.2\frac{ft}{s}$$

$$D_{FC} = 532.3 \times 2000 \times 0.02 + 0.05 \times \frac{200000^2}{532.3 \times 2000} = 23170\ lb$$

$$sin\gamma_{FC} = 0.31 - \frac{23170}{200000} = 0.1941\ ;\quad \dot{h}_{FC} = 669.2 \times 0.1941 = 129.9\frac{ft}{s},$$

For *20,000ft*,

$$\rho = 0.001267\frac{slug}{ft^3}\ ;\ \frac{T}{W} = 0.25 \times \frac{0.001267}{0.002377} = 0.1652$$

$$E_M = \sqrt{\frac{1}{4KC_{D0}}} = 15.81\ ;\ \frac{T}{S} = \frac{T}{W}\frac{W}{S} = 0.1652 \times 100 = 16.52\frac{lb}{ft^2}$$

$$Q_{FC} = \frac{16.52}{6 \times 0.02}\left[1 + \sqrt{1 + \frac{3}{\{15.81 \times 16.52\}^2}}\right] = 302.9\frac{lb}{ft^2}$$

$$V_{FC} = \sqrt{\frac{2 \times 302.9}{0.001267}} = 691.48\frac{ft}{s}$$

$$D_{FC} = 302.9 \times 2000 \times 0.02 + 0.05 \times \frac{200000^2}{302.9 \times 2000} = 15420\ lb$$

$$sin\gamma_{FC} = 0.1652 - \frac{15420}{200000} = 0.08815$$

$$\dot{h}_{FC} = 691.48 \times 0.08815 = 60.95\frac{ft}{s}$$

Observations

- *Airspeed corresponding to fastest climb increases with altitude*
- *Climb angle corresponding to fastest climb decreases with altitude*
- *Fastest rate of climb decreases with altitude*

Example 4.6
Draw a plot of airspeed corresponding to fastest rate of climb versus altitude for the airplane of Example 4.5.

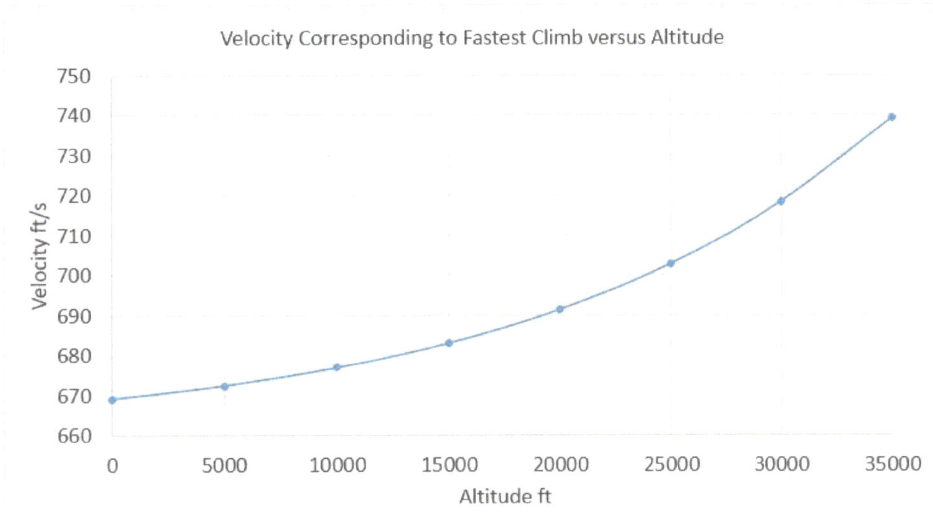

Figure 4.5: Airspeed corresponding to the fastest rate of climb versus altitude.

Example 4.7
Draw a plot of fastest rate of climb versus altitude for the airplane of Example 4.5.

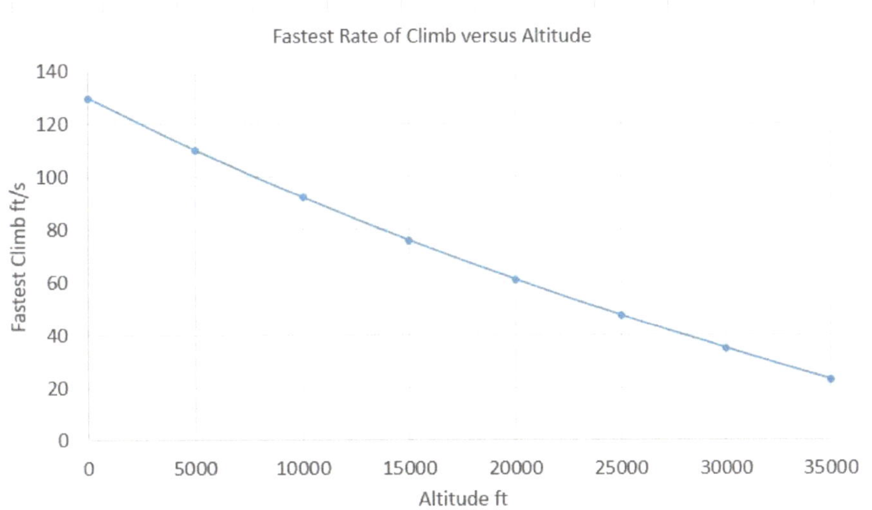

Figure 4.6: The fastest rate of climb versus altitude.

4.3 Fuel Consumption in Fastest Climb

We want to calculate the fuel consumption in fastest climb from an altitude of H_1 to H_2. First, we calculate the time required in the climb by integrating Eqn. (4.28) as

$$t = \int_{H_1}^{H_2} \frac{dh}{V_{FC} sin\gamma_{FC}}$$

This integration is complex because the airspeed and the climb angle depend on the altitude. Next, we calculate the fuel consumption by integrating Eqn. (3.17)

$$W_f = \int_{W_2}^{W_1} dW = \int_0^t cTdt = c\int_0^t \left(\frac{T}{W}\right) Wdt$$

This integration is complex because thrust to weight ratio is a function of time through change in altitude. Thrust specific fuel consumption c is a constant for a given engine.

We will demonstrate an approximation to estimate the amount of fuel consumption. In this method, we partition the altitude change H_1 to H_2 into several intervals. For each interval, we write the integrals as

$$\Delta t = \frac{\Delta h}{\dot{h}_{average}} \ ; \ \Delta W_f = c\left(\frac{T}{W}\right)_{average} W\Delta t$$

Readers who are familiar with the methods of numerical integration will recognize this procedure as the trapezoidal rule.

We will also use the following approximations:

- Ignore the dependence of thrust from the engine on the airspeed (see Section 8.6).
- Ignore the change in the weight of the airplane due to fuel consumption. We will show that this is a reasonable approximation because weight of fuel burnt during climb is small in comparison to the weight of the airplane.

<u>*Example 4.8*</u>

A 200000lb airplane has a wing area of 2000ft^2, C_{D0} = 0.02, K = 0.05, thrust to weight ratio at sea level of 0.31. Determine the fuel consumption for fastest rate of climb from sea level to 35000ft. Thrust specific fuel consumption c = 0.4lb/h.lb = (0.4/3600)lb/s.lb.

We partition the climb into 5000ft thick layers. We follow the procedure of Example 4.5 to calculate the rate of fastest climb and the thrust to weight ratio at various altitudes. We have tabulated the results in Figure 4.7.

Altitude (ft)	Fastest Rate of Climb (ft/s)	T/W	Δt (s)	ΔWf (lb)
0	129.90	0.3100		
5000	110.10	0.2671	41.67	267.18
10000	92.14	0.2290	98.89	545.12
15000	75.80	0.1951	178.64	841.77
20000	60.95	0.1652	292.50	1170.99
25000	47.35	0.1390	461.68	1560.48
30000	34.80	0.1162	730.38	2071.03
35000	23.02	0.0963	1210.67	2858.13

Figure 4.7: Calculation of fuel consumption

Interval: 0 ft to 5000 ft

$$\Delta t = \frac{5000}{\frac{1}{2} \times (129.9 + 110.1)} = 41.67\ s$$

$$\Delta W_f = \frac{0.4}{3600} \times \left\{\frac{1}{2} \times (0.31 + 0.2671)\right\} \times 200000 \times 41.67 = 267.2lb$$

Interval: 5000ft to 10000ft

$$\Delta t = \frac{5000}{\frac{1}{2} \times (110.1 + 92.14)} = 98.89\ s$$

$$\Delta W_f = \frac{0.4}{3600} \times \left\{\frac{1}{2} \times (0.2671 + 0.2290)\right\} \times 200000 \times 98.89 = 545.1lb$$

In the fourth and the fifth columns of Figure 4.7, we show the time intervals and thrust to weight ratio in seven *5000ft* layers. We can calculate the net fuel consumption for a climb to *35000ft* as *9315lb* by adding all the fuel weights in the fifth column. The fuel weight for the climb is *4.66%* of the weight of the airplane.

5 Landing

5.1 Introduction

The landing operation consists of three parts: approach, flare, and ground run. Large long-haul airplanes need longer ground run compared to the small short-hop airplanes. In addition, length of ground run increases with the altitude of the airport. Length and altitude of the runway at the airports, that we want our airplane to serve, affect our design decisions, such as, the wing loading and the high lift devices in the airplane.

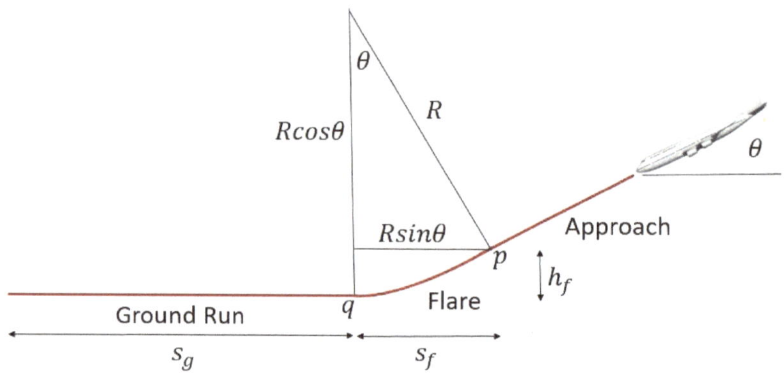

Figure 5.1: Landing phases

The flight path during approach makes an angle θ with the horizontal direction. The approach angle is typically $2^o < \theta < 4^o$. Flare is a nearly circular path that begins at p and ends at q. At q, the direction of motion of the airplane is horizontal along the runway. The airplane comes to a stop after the ground run. The airplane needs a runway length that is equal to the sum of the flare length s_f and ground run s_g .

In our analysis, we will assume that velocities at p and q are equal and are equal to the landing speed.

$$V_p = V_q = V_{LA}$$

5.2 Approach/Descent

The free body diagram for approach is

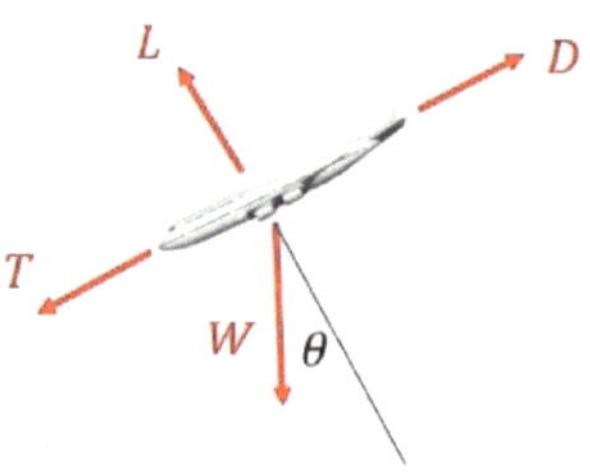

Figure 5.2: Free-body-diagram for descent.

For an approach with constant airspeed, the forces are in equilibrium. For a small angle of approach, $cos\theta \approx 1$ and the equations of equilibrium are

$$L = W_{LA}cos\theta \approx W_{LA} \tag{5.1}$$

$$T + W_{LA}sin\theta - D = 0 \tag{5.2}$$

From Eqn. (5.1), we can find the stall airspeed in terms of the density, landing weight, wing area, and maximum lift coefficient for landing (see Example 2.20).

$$V_{stall} = \sqrt{\left(\frac{2}{\rho C_{L,max,LA}}\right)\left(\frac{W_{LA}}{S}\right)} \tag{5.3}$$

Observations:

- Stall airspeed is larger at higher altitude.
- Stall airspeed is smaller for smaller wing loading. However, reducing the wing loading reduces the best range airspeed and consequently the best range (see Eqn. 3.33).
- Stall airspeed decreases with increase in maximum lift coefficient at landing.

Typically, the landing airspeed is 15% more than the stall airspeed.

$$V_{LA} = 1.15V_{stall} \tag{5.4}$$

The dynamic pressure corresponding the landing speed is

$$Q_{LA} = \frac{1}{2}\rho V_{LA}^2 \tag{5.5}$$

The drag is

$$D_{LA} = Q_{LA}S(C_{D0} + KC_{L,max,LA}^2) = Q_{LA}SC_{D0} + K\frac{W_{LA}^2}{Q_{LA}S} \tag{5.6}$$

From Eqn. (5.2),

$$\frac{T_{LA}}{W_{LA}} = \frac{D_{LA}}{W_{LA}} - sin\theta = \frac{Q_{LA}C_{D0}}{\left(W_{LA}/_S\right)} + \left(\frac{K}{Q_{LA}}\right)\left(\frac{W_{LA}}{S}\right) - sin\theta \tag{5.7}$$

<u>*Example 5.1*</u>

An airplane has wing loading of 120lb/ft², maximum lift coefficient at landing of 2.1, C_{D0} = 0.03, and K = 0.05. Determine the landing airspeed and thrust to weight ratio at landing at an altitude of 5000ft (ρ = 0.002048slug/ft³). The angle of approach is 3°.

From Eqns. (5.3) and (5.4),

$$V_{stall} = \sqrt{\left(\frac{2 \times 120}{0.002048 \times 2.1}\right)} = 236.2\frac{ft}{s} \quad ; \quad V_{LA} = 1.15 \times 236.2 = 271.7\frac{ft}{s}$$

From Eqn. (5.5)

$$Q_{LA} = \frac{1}{2} \times 0.002047 \times 271.7^2 = 75.57 \frac{lb}{ft^2}$$

From Eqn. (5.7)

$$\frac{T_{LA}}{W_{LA}} = \frac{75.57 \times 0.03}{120} + \left(\frac{0.05}{75.57}\right) \times 120 - 0.05236 = 0.04595$$

5.3 Flare

In the near circular path of the flare, the centripetal acceleration is

$$a = \frac{V_{LA}^2}{R} \leq 1.2g$$

Typically, we maximum centripetal acceleration we allow is $1.2g$. Therefore,

$$R \geq \frac{V_{LA}^2}{1.2g}$$

The flare height and length are

$$h_f = R(1 - cos\theta) \ ; \ \ s_f = Rsin\theta$$

5.4 Ground Run

The free body diagram for the ground run is

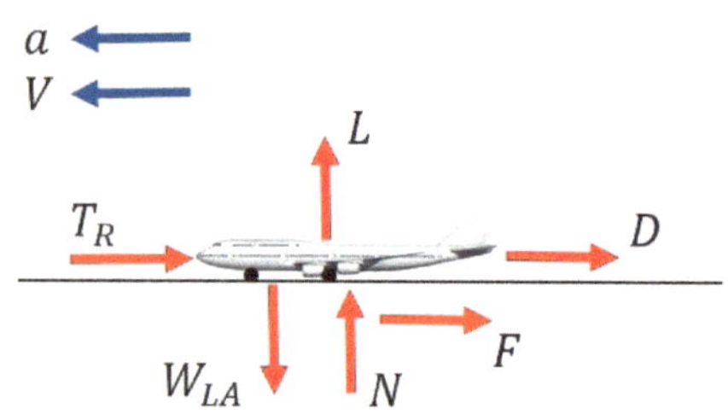

Figure 5.3: Free-body-diagram for landing ground run

The analysis for ground run in landing is similar to that for takeoff (see Section 4.2). The only difference is that we replace thrust T by reversed thrust T_R. In turbojet engines, two buckets turn the exhaust to produce reverse thrust.

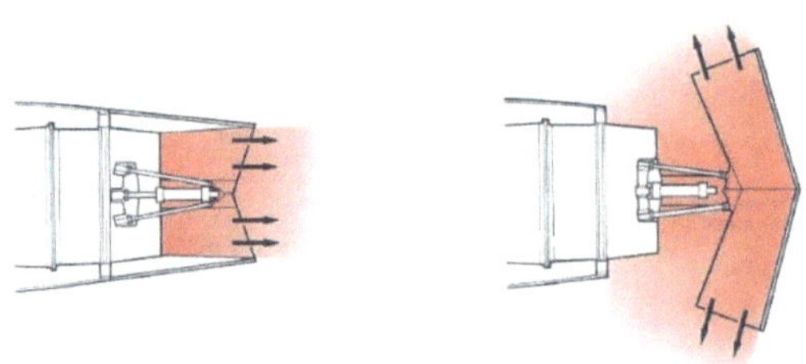

Figure 5.4: Thrust reversal in turbojet (diagram from Purdue AAE Propulsion)

In turbofan engines, openings turn the cold stream to produce reverse thrust.

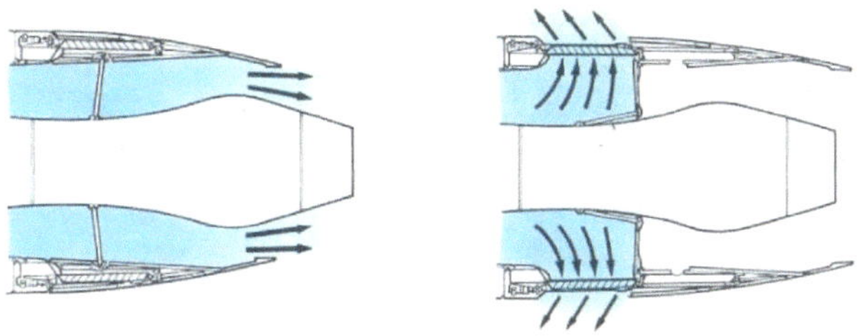

Figure 5.5: Thrust reversal in turbofan (diagram from Purdue AAE Propulsion)

Approximations:

- The reverse thrust is 40% or 50% of the maximum forward thrust (Raymer, 2006, pg. 551-552) and is constant during the entire ground run.
- We will ignore the increased drag from the spoilers.

Figure 5.6: Spoilers

Thus, from Eqn. (4.1)

$$\frac{g}{W_{LA}}(-T_R - \mu W_{LA}) + \frac{1}{2}\frac{g}{W_{LA}}\rho V^2 S(\mu C_L - C_D) = \frac{dV}{dt} \quad (5.8)$$

We define

$$A = \frac{g}{W_{LA}}(-T_R - \mu W_{LA}) \ ; \ B = \frac{1}{2}\frac{g}{W_{LA}}\rho S(\mu C_L - C_D) \quad (5.9)$$

By combining Eqns. (5.8) and (5.9),

$$A + BV^2 = \frac{dV}{dt} = \frac{dV}{dx}\frac{dx}{dt} = V\frac{dV}{dx} \quad (5.10)$$

By rearranging Eqn. (5.10),

$$dx = \frac{VdV}{A + BV^2} \quad (5.11)$$

By using Eqn. (4.7),

$$s_g = \int_{LA}^{stop} dx = \frac{1}{2B}\int_{LA}^{stop}\frac{dz}{z} = \frac{1}{2B}ln\left(\frac{z_{stop}}{z_{LA}}\right) \quad (5.12)$$

At landing, $V = V_{LA}$, and at stop, $V = 0$. Thus

$$s_g = \frac{1}{2B} ln\left(\frac{A}{A + BV_{LA}^2}\right) = -\frac{1}{2B} ln\left(1 + \frac{BV_{LA}^2}{A}\right) \tag{5.13}$$

From Eqn. (4.9),

$$s_g = -\frac{1}{2B}\frac{BV_{LA}^2}{A} = -\frac{V_{LA}^2}{2A} = \frac{W_{LA}}{2g}\frac{V_{LA}^2}{(T_R + \mu W_{LA})} = \frac{V_{LA}^2}{2g\left\{\left({T_R}/{W_{LA}}\right) + \mu\right\}} \tag{5.14}$$

By combining Eqns. (5.3), (5.4), and (5.14),

$$s_g = \frac{1.323\left({W_{LA}}/{S}\right)}{\rho g C_{L,max,LA}\left\{\left({T_R}/{W_{LA}}\right) + \mu\right\}} \tag{5.15}$$

Notes for Using Eqn. (5.15)

- *Reverse thrust is about 40% of the maximum thrust. We also correct the reverse thrust for altitude by using Eqn. (8.19).*
- *Landing weight W_{LA} and takeoff weight W_{TO} are not equal. We can relate these two weights by using Eqn. (3.19).*

$$\zeta = 1 - \frac{W_{LA}}{W_{TO}}$$

- *We will ignore the effect of braking, and use the rolling friction from Figure 4.2.*

Example 5.2

An airplane has wing loading of 120lb/ft², maximum lift coefficient at landing of 2.1, reverse thrust to weight ratio of 0.08, rolling friction coefficient of 0.03. Determine the ground run at landing at an altitude of 5000ft (ρ = 0.002048slug/ft³).

From Eqn. (5.14),

$$s_g = \frac{1.323 \times 120}{0.002048 \times 32.2 \times 2.1 \times \{0.08 + 0.03\}} = 10240\ ft$$

Example 5.3
An airplane has a maximum lift coefficient at landing of 2.1, reverse thrust to weight ratio of 0.08, rolling friction coefficient of 0.03. Determine the wing loading for a ground run at landing of 8500ft at an altitude of 5000ft (ρ = 0.002048slug/ft^3).

From Eqn. (5.14),

$$8500 = \frac{1.323 \times \left(W_{LA}/_S\right)}{0.0020148 \times 32.2 \times 2.1 \times \{0.08 + 0.03\}}$$

$$\left(W_{LA}/_S\right) = 97.87\frac{lb}{ft^2}$$

Example 5.4
An airplane has a takeoff weight of 800000lb, maximum lift coefficient at landing of 2.1, C_{D0} = 0.03, K = 0.05, reverse thrust to weight ratio of 0.08, rolling friction coefficient of 0.03. At the end of the cruise, the "cruise fuel weight fraction" is 0.3. Determine the wing area for a ground run at landing of 8500ft at an altitude of 5000ft (ρ = 0.002048slug/ft^3).

From Example 5.3,

$$\left(W_{LA}/_S\right) = 97.87\frac{lb}{ft^2}$$

From Eqn. (3.19),

$$0.3 = 1 - \frac{W_{LA}}{800000} \; ; \; W_{LA} = 560000\ lb$$

$$S = \frac{560000}{97.87} = 5722\ ft^2$$

5.5 Length of Runway

The pilot turns on the reverse thrust after the wheels touch down. Usually, there is a pilot response time, t_{pilot}, during which the airplane rolls freely with the landing airspeed without thrust reversal. Typically, $t_{pilot} = 3s$ and the net runway length for landing is

$$s = s_g + s_f + 3V_{LA}$$

5.6 Unpowered Descent

Unpowered descent is a situation when an airplane has lost its power or for a glider. We set thrust equal to zero in the equilibrium equation (5.2). Thus,

$$L = W \; ; \; sin\theta = \frac{D}{W} = \frac{D}{L} = \frac{C_D}{C_L} = \frac{1}{E} \tag{5.16}$$

For slowest descent, $sin\theta$ is minimum and E is maximum. By using Eqn. (2.41),

$$sin\theta_{min} = \frac{1}{E_M} = \sqrt{4KC_{DO}} \qquad (5.17)$$

From Eqn. (2.40), the lift coefficient for slowest descent is

$$C_{L,EM} = \sqrt{\frac{C_{DO}}{K}} \qquad (5.18)$$

Airspeed corresponding to slowest descent is

$$V_{slow} = \sqrt{\frac{2(W/S)}{\rho C_{L,EM}}} \qquad (5.19)$$

The airspeed is a function of altitude because density is a function of altitude. We can calculate the time the airplane remains airborne as it loses altitude from h_1 to h_2 by integrating the equation

$$\frac{dh}{dt} = -V_{slow} sin\theta_{min} = -\frac{V_{slow}}{E_M}$$

By using Eqns. (5.19) and (1.1)

$$t = -E_M \int_{h_1}^{h_2} \frac{dh}{V_{slow}} = -E_M \sqrt{\frac{\rho_{SL} C_{L,EM}}{2\left(\frac{W}{S}\right)}} \int_{h_1}^{h_2} e^{-(h/66666)} dh$$

By integrating,

$$t = 66666 E_M \sqrt{\frac{\rho_{SL} C_{L,EM}}{2\left(\frac{W}{S}\right)}} \left\{ e^{-(h_2/66666)} - e^{-(h_1/66666)} \right\} \qquad (5.20)$$

We can find the horizontal distance the airplane flies as it loses altitude from h_1 to h_2.

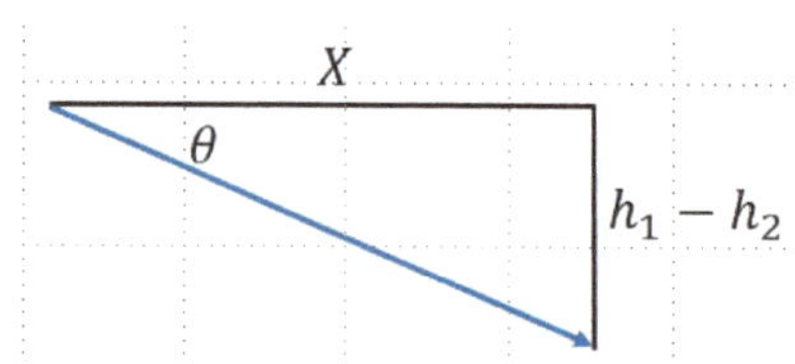

For small angle of descent

$$\frac{h_1 - h_2}{X} = tan\theta = sin\theta = \frac{1}{E_M} \;\; ; \;\; X = E_M(h_1 - h_2)$$

Example 5.5

An airplane has a wing loading of 120lb/ft², C_{D0} = 0.02, and K = 0.06. In an unpowered flight, determine the horizontal distance traveled by the airplane and the time interval as it descends from 25000ft to 5000ft.

$$E_M = \frac{1}{\sqrt{4 \times 0.06 \times 0.02}} = 14.43$$

$$X = 14.43 \times (25000 - 5000) = 288700\ ft = 55\ miles$$

$$C_{L,EM} = \sqrt{\frac{0.02}{0.06}} = 0.5774$$

$$t = 66666 \times 14.43 \sqrt{\frac{0.002377 \times 0.5774}{2 \times 120}} \left\{ e^{-\left(5000/66666\right)} - e^{-\left(25000/66666\right)} \right\}$$

$$= 553.3\ s$$

6 Turning Flight

6.1 Introduction

An essential maneuver of an airplane is a bank turn that changes the heading of the airplane. A bank is rolling the airplane to the right or left. The pilot uses two control surfaces, ailerons and spoilers to initiate a banking turn.

A down aileron increases the lift and an up aileron reduces the lift. By increasing the lift from the right wing and decreasing the lift from the left wing, the pilot makes the left wing dip and the airplane banks left. The overall effect of banking is a tilted lift vector. The vertical component of the lift balances the weight of the airplane and the horizontal component produces the centripetal acceleration, which is perpendicular to the flight path, necessary for the turn.

Ailerons are hinged flight control surfaces that are a part of the trailing edge of the wing. Down aileron increases the camber of the wing resulting in an increase in lift.

Figure 6.1: Ailerons

The spoilers are plates on the top surface of a wing. The pilot can reduce the lift from one wing by lifting the spoiler upward on that wing.

Figure 6.2: Spoilers

When the lift from one wing increases at constant density and airspeed the lift coefficient and the induced drag increases. The larger drag from one wing tends to move the nose of the airplane away from the flight path. The pilot uses the rudder to keep the nose pointed along the flight path to coordinate the turn.

In a banking turn, the lift and induced drag are larger than the lift and induced drag for level flight. The larger drag causes a decrease in airspeed. To maintain the airspeed and to prevent stall in a turn, the pilot increases the thrust to balance the increased drag. In this chapter, we will focus on the thrust and lift requirements during a banking turn and determine the requirements for the tightest turn with smallest radius of turn.

6.2 Equations of Motion

We write the equations of motion of the airplane for constant airspeed from two views of the airplane.

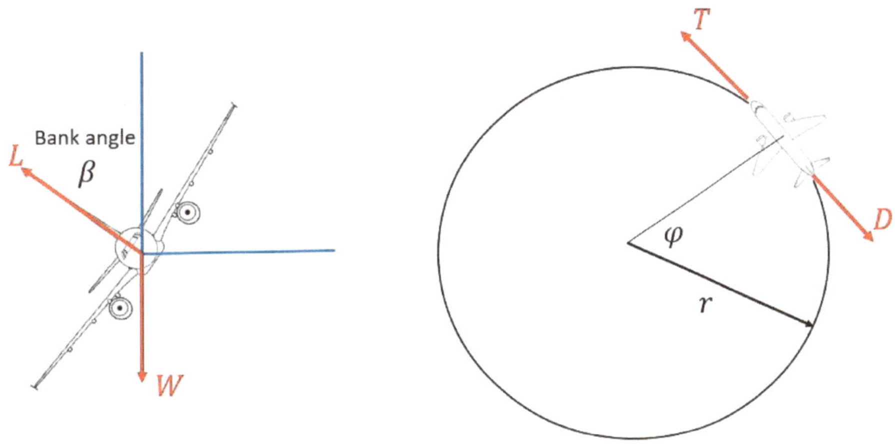

Figure 6.3: Turning flight

$$L cos\beta = W \tag{6.1}$$

$$L sin\beta = \frac{W}{g}\frac{V^2}{r} = \frac{W}{g}V\frac{d\varphi}{dt} \tag{6.2}$$

$$T = D = QS(C_{D0} + KC_L^2) \tag{6.3}$$

6.2 Load Factor

We define the load factor as (see Section 2.24)

$$n = \frac{L}{W} = \frac{L}{D}\frac{D}{W} = E\frac{T}{W}$$

The minimum load factor is one for level flight with zero bank angle. The maximum load factor for a given thrust to weight ratio is

$$n_{max} = E_M\left(\frac{T}{W}\right) \tag{6.4}$$

Observations

- The maximum load factor is not necessarily the operational limit of the load factor.
- We determine the operational limit by avoiding stall and by ensuring structural integrity (see Section 2.24).

From Eqn. (6.1), the maximum bank angle is

$$\beta_{max} = cos^{-1}\left(\frac{W}{L}\right) = cos^{-1}\left(\frac{1}{n_{max}}\right)$$

Higher bank angle requires larger lift. However, there is an upper limit of lift at stall.

6.3 Stall in Turning Flight

By using the definition of load factor,

$$L = \frac{1}{2}\rho V^2 S C_L = nW\ ;\ \ C_L = \frac{nW}{QS}$$

At stall in turning flight,

$$V_{stall,turn} = \sqrt{\frac{2n\left(W/_S\right)}{\rho C_{L,max}}} \tag{6.5}$$

A stall in level flight corresponds to $n=1$, thus

$$V_{stall,level} = \sqrt{\frac{2\left(W/_S\right)}{\rho C_{L,max}}} \tag{6.6}$$

The relationship between these two stall velocities is

$$V_{stall,turn} = n^{\frac{1}{2}} V_{stall,level} \tag{6.7}$$

Observation

- In turning flight, the load factor is more than one. Stall airspeed in turning flight is larger than the stall airspeed in level flight. Thus, an airplane may stall in a turning flight at a seemingly safe airspeed.

6.4 Airspeed in Turning Flight

From Eqn. (6.3),

$$T = QS\{C_{DO} + KC_L^2\} = QS\left\{C_{DO} + K\frac{n^2W^2}{Q^2S^2}\right\} \tag{6.8}$$

By rearranging,

$$C_{DO}S^2Q^2 - TSQ + Kn^2W^2 = 0$$

By solving the quadratic equation for Q,

$$Q = \frac{TS \pm \sqrt{T^2S^2 - 4KC_{DO}n^2S^2W^2}}{2C_{DO}S^2}$$

By using the definition E_M from Eqn. (2.41),

$$Q = \frac{\left(\frac{T}{S}\right)}{2C_{DO}}\left[1 \pm \sqrt{1 - \left\{\frac{n}{E_M\left(\frac{T}{W}\right)}\right\}^2}\,\right] = \frac{1}{2}\rho V^2 \tag{6.9}$$

$$V = \left[\frac{\left(\frac{T}{S}\right)}{\rho C_{DO}}\left[1 \pm \sqrt{1 - \left\{\frac{n}{E_M\left(\frac{T}{W}\right)}\right\}^2}\,\right]\right]^{\frac{1}{2}} \tag{6.10}$$

When the load factor is one, Eqn. (6.10) is identical to the Eqn. (3.11b) and yields the airspeed for level flight.

<u>Example 6.1</u>
An airplane has $E_M = 15$ and is being flown at a thrust to weight ratio fixed at 0.1, wing loading of 100lb/ft^2, at an altitude of 10000ft, and with $C_{DO} = 0.015$. Plot the airspeed as a function of the load factor. Also plot the stall airspeed for $C_{L,max} = 0.7$ by using Eqn. (6.5). ($\rho = 0.001756$ slug/ft^3)

For this flight,

$$n_{max} = E_M\left(\frac{T}{W}\right) = 1.5\ ;\ \frac{T}{S} = \frac{T}{W}\frac{W}{S} = 10\frac{lb}{ft^2}$$

We choose a set of values for the load factor between 1 and 1.5. For each value, we calculate two values of airspeed from Eqn. (6.10). We plot load factor along the abscissa and airspeed along the ordinate. The curve has an upper branch and a lower branch.

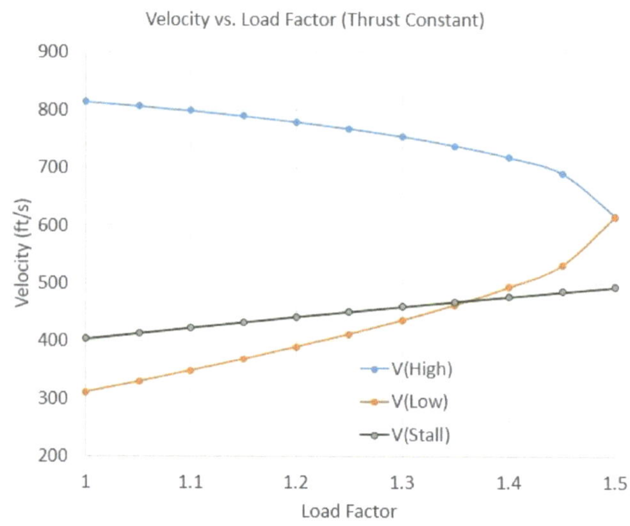

Figure 6.2: Airspeed vs. load factor

Observations

- The range of the load factor is $1 \leq n \leq 1.5$. Below the value of one, flight is not possible (see Section 3.2.1)
- The airspeed plot has two branches, one for the high airspeed corresponding to the positive sign in Eqn. (6.10) and the other for the low airspeed corresponding to the negative sign in Eqn. (6.10). The high airspeed branch is the stable branch (see Section 3.3).
- For turning flight, the load factor is more than one and for constant thrust, airspeed decreases with increase in load factor. For this constant thrust flight, the high airspeed is about *800ft/s* for level flight and decreases to about *600ft/s* at the maximum load factor.
- Consider an airplane in level flight ($n = 1$, $V = 800ft/s$) enters a turn ($n > 1$). If the pilot keeps the thrust unchanged, the airspeed in the turn will decrease. The pilot has to increase thrust to keep the airspeed unchanged.
- The $V_{stall,turn}$ is plotted against the load factor by using Eqn. (6.5).
- By comparing the stall airspeed with the airspeed in the low-airspeed branch, we see that a flight in the low-airspeed branch is impossible due to stall when the load factor is less than *1.35*.
- By comparing the stall airspeed with the airspeed in the high-airspeed branch, we see that a flight in the high-airspeed branch is possible for the entire range of the load factor.

6.4 Lift Coefficient in Turning Flight

From the definition of the load factor,

$$C_L = \frac{n\left(\frac{W}{S}\right)}{Q} \tag{6.11}$$

To prevent stall, we must satisfy the condition that

$$\frac{n\left(\frac{W}{S}\right)}{Q} \le C_{L,max} \tag{6.12}$$

<u>*Example 6.2*</u>

An airplane has a wing loading of 100lb/ft², $E_M = 15$, and $C_{D0} = 0.03$. We wish to turn this airplane with a bank angle of 30° with a thrust to weight ratio of 0.1. Is this turn possible when $C_{L,max} = 0.6$?

From Eqn. (6.1),

$$n = \frac{L}{W} = \frac{1}{cos30^o} = 1.155$$

$$\frac{T}{S} = \frac{T}{W}\frac{W}{S} = 0.1 \times 100 = 10\frac{lb}{ft^2}$$

From Eqn. (6.9),

$$Q = \frac{10}{2 \times 0.03}\left[1 \pm \sqrt{1 - \left\{\frac{1.155}{15 \times 0.1}\right\}^2}\right] = 273.0 \; and \; 60.33\frac{lb}{ft^2}$$

From Eqn. (6.11)

$$C_L = \frac{n\left(\frac{W}{S}\right)}{Q} = \frac{1.155 \times 100}{Q} = 0.4231 \; and \; 1.9146$$

A turn with 273lb/ft² of dynamic pressure is possible. A turn with 60.33lb/ft² of dynamic pressure is not possible.

<u>*Example 6.3*</u>

An airplane has a wing loading of 100lb/ft², $C_{D0} = 0.015$, and $E_M = 15$. We wish to turn this airplane with a bank angle of 30° with a thrust to weight ratio of 0.08. Is this turn possible when $C_{L,max} = 0.6$?

From Eqn. (6.1),

$$n = \frac{L}{W} = \frac{1}{cos30^o} = 1.155$$

$$\frac{T}{S} = \frac{T}{W}\frac{W}{S} = 0.08 \times 100 = 8\frac{lb}{ft^2}$$

From Eqn. (6.9),

$$Q = \frac{8}{2 \times 0.03}\left[1 \pm \sqrt{1 - \left\{\frac{1.155}{15.81 \times 0.08}\right\}^2}\right] = 169.5 \; and \; 97.16\frac{lb}{ft^2}$$

From Eqn. (6.11)

$$C_L = \frac{n\left(\frac{W}{S}\right)}{Q} = \frac{1.155 \times 130}{Q} = 0.6814 \; and \; 1.189$$

This turn is not possible because of insufficient thrust.

Observation

- For turning flight, airspeed decreases as the airplane enters a turn. To maintain the vertical equilibrium and prevent any loss of altitude, the pilot has to increase the airspeed by increasing thrust.

6.5 Thrust

We can calculate the required thrust for a turn at a given bank angle. From Eqn. (6.3),

$$T = D = QS(C_{D0} + KC_L^2)$$

By using Eqn. (6.11),

$$\frac{T}{S} = Q\left[C_{D0} + K\left\{\frac{n\left(W/_S\right)}{Q}\right\}^2\right] \tag{6.13a}$$

$$n^2 = \frac{\left(\frac{T}{W}\right)\left(\frac{W}{S}\right)Q - Q^2 C_{D0}}{K\left(\frac{W}{S}\right)^2} \tag{6.13b}$$

Example 6.4

An airplane with a wing loading of 100lb/ft², C_{D0} = 0.015, K = 0.05 is in level flight with an airspeed of 550ft/s at sea level. The pilot wants to enter a turn with bank angle of 30°. Determine the required thrust to weight ratio for the turn. Ensure that the lift coefficient is less than 0.6.

$$\rho = 0.002377\, slug/ft^3 \quad ; \quad Q = 0.5 \times 0.002377 \times 550^2 = 359.5 \frac{lb}{ft^2}$$

$$n = \frac{1}{cos30^o} = 1.155$$

$$\frac{T}{S} = 359.5 \times \left[0.015 + 0.05 \times \left\{\frac{1.155 \times 100}{359.5}\right\}^2\right] = 7.248 \frac{lb}{ft^2}$$

$$\frac{T}{W} = \frac{T}{S}\frac{S}{W} = \frac{7.248}{100} = 0.07248$$

$$C_L = \frac{n\left(\frac{W}{S}\right)}{Q} = \frac{1.155 \times 100}{359.5} = 0.3213$$

Example 6.5

An airplane with a wing loading of 100lb/ft², $C_{D0} = 0.015$, $K = 0.05$ is in level flight with an airspeed of 750ft/s at an altitude of 30000ft. The pilot wants to enter a turn with bank angle of 30°. Determine the required thrust to weight ratio for the turn. Ensure that the lift coefficient is less than 0.6.

$$\rho = 0.0008907\, slug/ft^3 \quad ; \quad Q = 0.5 \times 0.001496 \times 750^2 = 250.5 \frac{lb}{ft^2}$$

$$n = \frac{1}{cos30^o} = 1.155$$

$$\frac{T}{S} = 250.5 \times \left[0.015 + 0.05 \times \left\{\frac{1.155 \times 100}{250.5}\right\}^2\right] = 6.420 \frac{lb}{ft^2}$$

$$\frac{T}{W} = \frac{T}{S}\frac{S}{W} = \frac{6.420}{100} = 0.0642$$

$$C_L = \frac{n\left(\frac{W}{S}\right)}{Q} = \frac{1.155 \times 100}{250.5} = 0.4611$$

6.6 Minimum Turning Radius

From Eqns. (6.1) and (6.2), the turning radius is

$$r = \frac{V^2}{g}\frac{1}{sin\beta}\frac{W}{L} = \frac{V^2}{g}\frac{cos\beta}{sin\beta} \tag{6.14}$$

$$\frac{cos\beta}{sin\beta} = \frac{1/n}{\sqrt{1 - 1/n^2}} = \frac{1}{\sqrt{n^2 - 1}}$$

$$r = \frac{V^2}{g}\frac{1}{\sqrt{n^2 - 1}} \tag{6.15}$$

For minimum turning radius,

$$\frac{dr}{dV} = \frac{2V}{g(n^2 - 1)^{\frac{1}{2}}} - \frac{1}{2}\frac{V^2}{g}\frac{2n}{(n^2 - 1)^{\frac{3}{2}}}\frac{dn}{dV} = 0$$

After simplification,

$$n^2 - 1 - \frac{1}{2}nV\frac{dn}{dV} = 0 \tag{6.16}$$

Also,

$$\frac{dn}{dV} = \frac{dn}{dQ}\frac{dQ}{dV} = \rho V\frac{dn}{dQ} = \frac{2Q}{V}\frac{dn}{dQ} \tag{6.17}$$

By combining Eqns. (6.16) and (6.17),

$$n^2 - 1 - nQ\frac{dn}{dQ} = 0 \tag{6.18}$$

By differentiating Eqn. (6.13b) with respect to Q ,

$$\frac{dn}{dQ} = \frac{(T/S) - 2QC_{D0}}{2Kn(W/S)^2} \tag{6.19}$$

By combining Eqns. (6.18) and (6.19),

$$n^2 = 1 + \frac{Q\left\{\left(\frac{T}{S}\right) - 2QC_{D0}\right\}}{2K(W/S)^2} \tag{6.20}$$

By combining Eqns. (6.13) and (6.20), we find the dynamic pressure for the tightest turn with minimum radius.

$$Q_{TT} = 2K\frac{(W/S)}{(T/W)} \tag{6.21}$$

From the definition of dynamic pressure, the airspeed for the tightest turn is

$$V_{TT} = \sqrt{\frac{2Q_{TT}}{\rho}} \tag{6.22}$$

From Eqn. (6.20), the load factor for the tightest turn is

$$n_{TT} = \sqrt{1 + \frac{Q_{TT}\left\{\left(\frac{T}{S}\right) - 2Q_{TT}C_{DO}\right\}}{2K\left(W/_S\right)^2}}$$

By substituting the dynamic pressure for the tightest turn from Eqn. (6.21),

$$n_{TT} = \sqrt{2\left[1 - \left\{\frac{2KC_{DO}}{\left(T/_W\right)^2}\right\}\right]} \tag{6.23}$$

We can find the banking angle for the tightest turn as

$$\beta_{TT} = cos^{-1}\left(\frac{1}{n_{TT}}\right) \tag{6.24}$$

From Eqn. (6.15), the radius of the tightest turn is

$$r_{TT} = \frac{V_{TT}^2}{g}\frac{1}{\sqrt{n_{TT}^2 - 1}} \tag{6.25}$$

From Eqn. (6.11), we can find the lift coefficient for the tightest turn. We must ensure that the airplane will not stall and the lift coefficient is less than the maximum lift coefficient. To prevent stall, the pilot may deploy flaps during a turn.

$$C_{L,TT} = \frac{n_{TT}\left(\frac{W}{S}\right)}{Q_{TT}} \leq C_{L,max} \tag{6.26}$$

Observations

- The dynamic pressure for tightest turn increases when the thrust to weight ratio decreases.
- The load factor decreases when the thrust to weight ratio decreases.
- There is no solution of Eqn. (6.23) when

$$\frac{T}{W} < \sqrt{2KC_{D0}}$$

- Only certain combinations of load factor and dynamic pressure satisfy Eqn. (6.26) and the airplane does not stall.
- A solution of Eqn. (6.24) for bank angle does not exist when the load factor is less than one.
- The pilot may decide to deploy flaps to increase the maximum lift coefficient. Thus, we have to use the proper maximum lift coefficient, with or without flaps, in Eqn. (6.26).
- For a fixed dynamic pressure, airspeed increases when density decreases. Consequently, the turning radius increases with altitude.

We can do this long calculation and analyze various "what if" situations by using the Excel spreadsheet (see Example E.1 in Appendix E).

<u>Example 6.6</u>

An airplane has a wing loading of 100lb/ft², $C_{D0} = 0.02$, $K = 0.05$. Determine the radius of the tightest turn at sea level. The maximum lift coefficient with flaps is 1.2. The maximum thrust to weight ratio is 0.25 at sea level.

For a real solution of Eqn. (6.23) for the load factor, we must have

$$1 - \left\{\frac{2KC_{D0}}{\left({}^{T}/_{W}\right)^2}\right\} > 0 \ or \ \frac{T}{W} > \sqrt{2KC_{D0}}$$

The minimum thrust to weight ratio is

$$\frac{T}{W} = \sqrt{2 \times 0.05 \times 0.02} = 0.04472$$

We will find the load factor and lift coefficient for a set of thrust to weight ratio to check the feasibility of a solution.

Setting	${}^{T}/_{W}$	n	C_L
1	*0.05*	*0.8944*	*0.4472*
2	*0.07*	*1.178*	*0.8246*
3	*0.09*	*1.277*	*1.149*
4	*0.11*	*1.324*	*1.456*

We discard Setting-1because the load factor is less than one and thrust is insufficient even for level flight. We discard Setting-4 because lift coefficient is larger than the maximum lift coefficient. Between Setting-2 and Setting-3, we choose Setting-3 because higher thrust to weight ratio gives smaller airspeed and smaller turning radius. For thrust to weight ratio of 0.09,

$$Q_{TT} = 2 \times 0.05 \times \frac{100}{0.09} = 111.1\, lb/ft^2 \;\; ; \; V_{TT} = \sqrt{\frac{2 \times 111.1}{0.002377}} = 305.8 \frac{ft}{s}$$

$$n_{TT} = \sqrt{2\left[1 - \left\{\frac{2 \times 0.05 \times 0.02}{(0.09)^2}\right\}\right]} = 1.277 \;\; ;$$

$$\beta_{TT} = cos^{-1}\left(\frac{1}{1.227}\right) = 0.6184\, rad = 38.43^o$$

$$r_{TT} = \frac{305.6^2}{32.2} \frac{1}{\sqrt{1.227^2 - 1}} = 3659\, ft = 0.693\, miles$$

7 Propeller Airplanes

7.1 Power

The difference between a jet engine and a propeller is that a jet engine supplies thrust and a propeller supplies power.

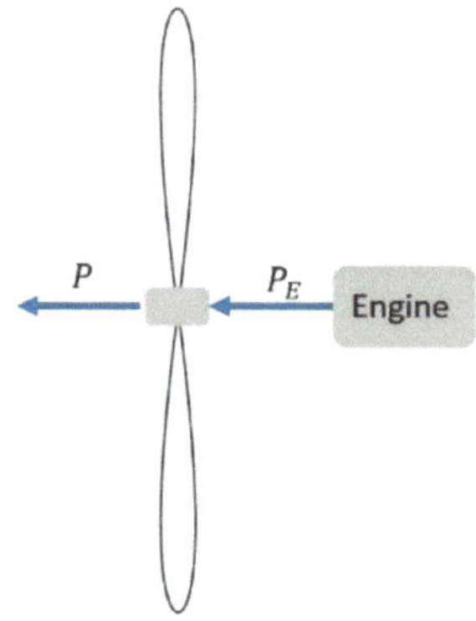

Figure 7.1: Engine and propeller

The engine supplies power P_E to the propeller and the propeller supplies power P to the airplane. We define the efficiency of the propeller as

$$\eta = \frac{P}{P_E} \tag{7.1}$$

The efficiency of a propeller is about 80% to 85%. We will discuss the characteristics of a propeller in Section 7.7. Manufacturers provide engine power in horsepower (HP). The conversion between *HP* and *ft.lb/s* is

$$1\ HP = 550\ ft.\frac{lb}{s}$$

Just like a jet engine (see Eqn. 8.19), the propeller power depends on the altitude

$$\frac{P}{P_{SL}} = \frac{\rho}{\rho_{SL}} \tag{7.2}$$

7.2 Level Flight

7.2.1 Cruise

The equations of equilibrium for cruise at constant airspeed are

$$T = D = QS(C_{D0} + KC_L^2) \tag{7.3}$$

$$W = L = QSC_L \tag{7.4}$$

By using Eqns. (7.3) and (7.4), we can write the expression for power-weight ratio as

$$\frac{P}{W} = \frac{TV}{W} = \frac{QVS(C_{D0} + KC_L^2)}{W} = \frac{\rho V^3 C_{D0}}{2(W/_S)} + \frac{2K(W/_S)}{\rho V} \tag{7.5}$$

Example 7.1

An airplane has wing loading of 60lb/ft², C_{D0} = 0.02, K = 0.05. Draw a power-weight ratio versus airspeed plot at sea level. (ρ = 0.002377slug/ft³)

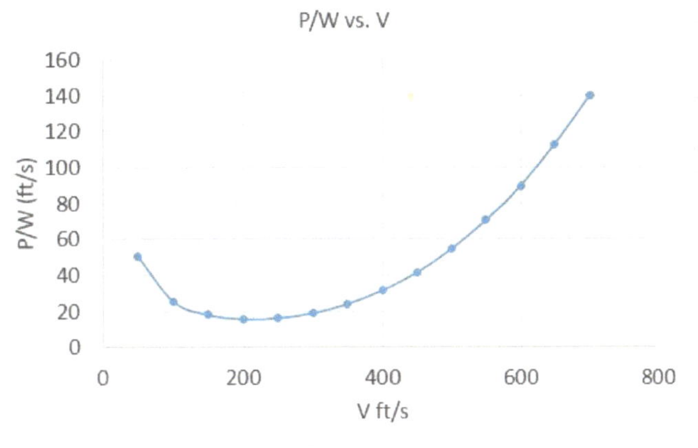

Figure 7.2: Example 7.1

Example 7.2

An airplane has wing loading of 60lb/ft², C_{D0} = 0.02, K = 0.05. For a sea level power-weight ratio of 55ft/s, determine the airspeed of level flight at an altitude of 10000ft. (ρ = 0.001756slug/ft³)

By using Eqn. (7.2), we calculate the power-weight ratio at 10000ft altitude

$$(P/_W)_{10000} = (P/_W)_{SL} \times \frac{0.001756}{0.002377} = 40.63\frac{ft}{s}$$

We solve Eqn. (7.5) graphically by plotting f(V).

$$f(V) = \frac{P}{W} - \frac{\rho V^3 C_{D0}}{2(W/_S)} - \frac{2K(W/_S)}{\rho V}$$

From Figure 7.3, airspeed is 490ft/s.

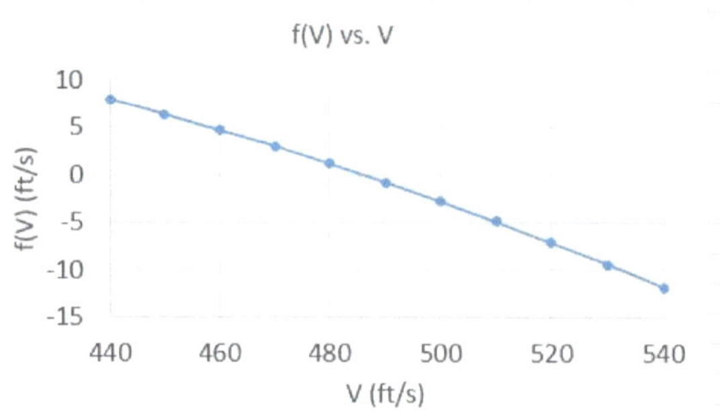

Figure 7.3: Example 7.2

7.2.2 Minimum Power

To find the airspeed that corresponds to the minimum power-weight ratio, we equate the derivative of the right-hand-side of Eqn. (7.5) with respect to V to zero.

$$\frac{3\rho V^2 C_{D0}}{2\left(W/_S\right)} - \frac{2K\left(W/_S\right)}{\rho V^2} = 0 \tag{7.6}$$

By solving Eqn. (7.6) for V

$$V_{Pmin} = \left\{\frac{2\left(W/_S\right)}{\rho}\right\}^{1/2} \left(\frac{K}{3C_{D0}}\right)^{1/4} \tag{7.7}$$

We find the corresponding lift coefficient from Eqn. (7.4),

$$\frac{W}{S} = \frac{1}{2}\rho V_{Pmin}^2 C_{L,Pmin} = \frac{W}{S}\left(\frac{K}{3C_{D0}}\right)^{1/2} C_{L,Pmin}$$

$$C_{L,Pmin} = \left(\frac{3C_{D0}}{K}\right)^{1/2} \tag{7.8}$$

$$C_{D,Pmin} = C_{D0} + KC_{L,Pmin}^2 = 4C_{D0} \tag{7.9}$$

$$E_{Pmin} = \frac{C_{L,Pmin}}{C_{D,Pmin}} = \left(\frac{3}{16KC_{D0}}\right)^{1/2} \tag{7.10}$$

By substituting Eqn. (7.7) in Eqn. (7.5),

$$\frac{P_{Min}}{W} = \frac{T_{Pmin}V_{Pmin}}{W} = \frac{D_{Pmin}V_{Pmin}}{L_{Pmin}} = \frac{C_{D,Pmin}V_{Pmin}}{C_{L,Pmin}} = \frac{V_{Pmin}}{E_{Pmin}} \tag{7.11}$$

7.2.3 Service Ceiling

At the service ceiling, the full throttle power-weight ratio is equal to the minimum, required power-weight ratio. Let the density at the service ceiling be $\rho_{Ceiling}$.

$$\left(\frac{P_{Max}}{W}\right)_{Ceiling} = \left(\frac{P_{Max}}{W}\right)_{SL} \frac{\rho_{Ceiling}}{\rho_{SL}} \tag{7.12}$$

From Eqn. (7.7),

$$V_{Pmin,Ceiling} = \left\{\frac{2\left(W/_S\right)}{\rho_{Ceiling}}\right\}^{1/2} \left(\frac{K}{3C_{D0}}\right)^{1/4} \tag{7.13}$$

From Eqn. (7.11),

$$\left(\frac{P_{Min}}{W}\right)_{Ceiling} = \frac{V_{Pmin,Ceiling}}{E_{Pmin}} \tag{7.14}$$

At the service ceiling

$$\left(\frac{P_{Max}}{W}\right)_{Ceiling} = \left(\frac{P_{Min}}{W}\right)_{Ceiling} \tag{7.15}$$

From Eqn. (7.15), we determine $\rho_{Ceiling}$ and determine the corresponding altitude from standard atmosphere table.

Example 7.3

In an airplane, wing loading is 40lb/ft², $C_{D0} = 0.025$, $K = 0.051$. Full throttle power-weight ratio at sea level is 55ft/s. Determine the service ceiling.

$$\left(\frac{P_{Max}}{W}\right)_{Ceiling} = 55 \times \frac{\rho_{Ceiling}}{0.002377} = 23140\rho_{Ceiling}\frac{ft}{s}$$

$$V_{Pmin,Ceiling} = \left\{\frac{2 \times 40}{\rho_{Ceiling}}\right\}^{1/2} \left(\frac{0.051}{3 \times 0.025}\right)^{1/4} = 8.122\left\{\frac{1}{\rho_{Ceiling}}\right\}^{1/2}\frac{ft}{s}$$

$$E_{Pmin} = \left(\frac{3}{16 \times 0.051 \times 0.025}\right)^{1/2} = 12.13$$

From Eqn. (7.14),

$$\left(\frac{P_{Min}}{W}\right)_{Ceiling} = \frac{8.122}{12.13}\left\{\frac{1}{\rho_{Ceiling}}\right\}^{1/2} = 0.6696\left\{\frac{1}{\rho_{Ceiling}}\right\}^{1/2}$$

From Eqn. (7.15),

$$23140\rho_{Ceiling} = 0.6696\left\{\frac{1}{\rho_{Ceiling}}\right\}^{1/2}$$

$$\rho_{Ceiling} = 0.0009425\frac{slug}{ft^3}$$

Altitude corresponding to this density is about 28500ft.

7.3 Range in Level Flight

For a propeller-engine combination, the weight of the fuel that the engine burns is proportional to the power supplied by the engine to the propeller, P_E. The constant of proportionality is the specific fuel consumption, c_P, of the engine. Thus

$$-\frac{dW}{dt} = c_P P_E = \frac{c_P}{\eta}P \tag{7.16}$$

The negative sign signifies that the weight of the airplane decreases as the engine burns fuel. With X being the distance traveled by the airplane in level flight, we write the airspeed of the airplane as

$$\frac{dX}{dt} = V \tag{7.17}$$

By dividing Eqn. (7.16) by Eqn. (7.17)

$$\frac{dX}{dW} = -\frac{\eta}{c_P}\frac{V}{P} = -\frac{\eta}{c_P}\frac{1}{T} = -\frac{\eta}{c_P}\frac{L}{D}\frac{1}{L} = -\frac{\eta}{c_P}\frac{C_L}{C_D}\frac{1}{L} = -\frac{\eta}{c_P}\frac{E}{W}$$

$$dX = -\frac{\eta E}{c_P}\frac{dW}{W} \tag{7.18}$$

7.3.1 Constant Lift Coefficient

By integrating Eqn. (7.18) between two locations on the path of the airplane in a level flight with constant lift coefficient (therefore, constant E) we get the range as

$$X_{CL} = -\frac{\eta E}{c_P}\ell n\left(\frac{W_2}{W_1}\right)$$

The weights of the airplane at the beginning and at the end of the cruise are W_1 and W_2. The relationship between these weights and the weight of consumed fuel, W_f is

$$W_2 = W_1 - W_f$$

Then

$$X_{CL} = -\frac{\eta E}{c_P} \ell n\left(\frac{W_1 - W_f}{W_1}\right) == -\frac{\eta E}{c_P} \ell n(1-\zeta) = \frac{\eta E}{c_P} \ell n\left(\frac{1}{1-\zeta}\right) \tag{7.19}$$

The "cruise fuel weight fraction" is ζ .

Observations:

- Airspeed indirectly appears in this range equation through C_L in E.
- For constant altitude (hence constant density) cruise with constant lift coefficient, the pilot must reduce the airspeed, as the airplane gets lighter due to fuel consumption.
- For constant airspeed cruise with constant lift coefficient, the pilot must use the cruise-climb flight program to reduce the density, as the airplane gets lighter due to fuel consumption.
- During the cruise, airspeed decreases in the constant altitude flight, therefore, the flight time for this program is longer than the flight time for the cruise-climb program.

For best range, we maximize the lift to drag ratio to E_M (see Eqns. (2.39), (2.40), (2.41))

$$X_{BR,CL} = \frac{\eta E_M}{c_P} \ell n\left(\frac{1}{1-\zeta}\right) \tag{7.20}$$

The lift coefficient during this flight is

$$C_{L,BR} = C_{L,EM} = \sqrt{\frac{C_{D0}}{K}} \tag{7.21}$$

For constant altitude flight, we can determine two "best range" velocities; one corresponds to the start of the flight when the airplane weighs W_1 and the other corresponds to the end of the flight when the airplane weighs W_2.

$$V_{BR,CL,1} = \sqrt{\frac{2\left({W_1}/{S}\right)}{\rho C_{L,EM}}} \tag{7.22}$$

$$V_{BR,CL,2} = \sqrt{\frac{2\left({W_2}/{S}\right)}{\rho C_{L,EM}}} = \sqrt{\frac{2\left({W_1}/{S}\right)(1-\zeta)}{\rho C_{L,EM}}} \tag{7.23}$$

The power-weight ratio at the beginning of the flight is larger than that at the end of the flight. We determine the larger power-weight ratio for best range by inserting the best-range conditions in Eqn. (7.5)

$$\frac{P_{BR}}{W} = \frac{TV_{BR,CL,1}}{W} = \frac{C_{D,BR}V_{BR,CL,1}}{C_{L,BR}} = \frac{V_{BR,CL,1}}{E_M} \tag{7.24}$$

<u>*Example 7.4*</u>

In an airplane, $W_1/S = 40\ lb/ft^2$, $C_{D0} = 0.02$, $K = 0.05$, $\eta = 0.85$, $c_P = 0.5\ lb/h.HP$. Determine the best range for constant lift coefficient, the starting and ending velocities for constant altitude of 10000ft, and the power to weight ratio when $\zeta = 0.2$. ($\rho = 0.001756\ slug/ft^3$)

From Eqn. (2.41)

$$E_M = \sqrt{\frac{1}{4KC_{D0}}} = 15.8$$

$$X_{BR,CL} = \frac{0.85 \times 15.8 \times 3600 \times 550}{0.5} \ell n\left(\frac{1}{1-0.2}\right) = 2248\ miles$$

$$C_{L,EM} = \sqrt{\frac{0.02}{0.05}} = 0.63$$

$$V_{BR,CL,1} = \sqrt{\frac{2 \times 40}{0.001756 \times 0.63}} = 268.9\ ^{ft}/_{s} = 183.4\ mph$$

$$V_{BR,CL,2} = \sqrt{\frac{2 \times 40 \times (1-0.2)}{0.001756 \times 0.63}} = 240.5\ ^{ft}/_{s} = 164\ mph$$

$$\frac{P_{BR}}{W} = \frac{268.9}{15.8} = 17.02\frac{ft}{s}$$

<u>*Example 7.5*</u>

In an airplane, $W_1/S = 40\ lb/ft^2$, $C_{D0} = 0.02$, $K = 0.05$, $\eta = 0.85$, $c_P = 0.5\ lb/h.HP$. Maximum power to weight ratio at sea level is 45ft/s. Determine the fuel consumption for a cruise of 1500 miles under "best range" conditions and at an altitude of 10000ft. Estimate the time of flight. ($\rho = 0.001756\ slug/ft^3$)

$$\left(\frac{P}{W}\right)_{10000} = 45 \times \frac{0.001756}{0.002377} = 33.24\frac{ft}{s}$$

$$c_P = 0.5\ \frac{lb}{h.HP} \times \frac{1HP}{550\ ^{lb.ft}/_{s}} \times \frac{1h}{3600s} = \frac{0.5}{550 \times 3600}\ per\ ft.$$

By inserting the value of E_M from Example 7.4 in Eqn. (7.20),

$$1500 \times 5280 = \frac{0.85 \times 15.8 \times 3600 \times 550}{0.5} \ell n\left(\frac{1}{1-\zeta}\right)$$

$$\ell n\left(\frac{1}{1-\zeta}\right) = 0.1489\ ;\ \zeta = 0.1383$$

From Example 7.4, the average airspeed is

$$V_{BR,CL,Av} = 0.5 \times (183 + 164) = 174\ mph$$

Time of flight is

$$\Delta t = \frac{1500}{174} = 8.62\ hours$$

From Example 7.4, the power to weight ratio is 17.02ft/s that is 51% of the maximum power to weight ratio.

7.3.2 Constant Altitude and Constant Airspeed

For this flight program, we divide Eqn. (7.17) by Eqn. (7.16) to get

$$\frac{dX}{dW} = -\frac{\eta}{c_P}\frac{V}{P} = -\frac{\eta}{c_P}\frac{1}{T} = -\frac{\eta}{c_P}\frac{1}{D}$$

$$dX = -\frac{\eta}{c_P}\frac{dW}{QS(C_{D0} + KC_L^2)} = -\frac{\eta}{c_P}\frac{dW}{QS\left\{C_{D0} + K\left(\frac{W}{QS}\right)^2\right\}}$$

$$dX = -\frac{\eta}{c_P}\frac{QS}{K}\left(\frac{dW}{\frac{Q^2S^2C_{D0}}{K} + W^2}\right)$$

As altitude is constant, density is constant. Since airspeed is also constant, the dynamic pressure Q is constant. By using the method of integration from Section 3.4.2, we can find the range as

$$X_{V,\rho} = \frac{\eta}{c_P}\frac{1}{\sqrt{KC_{D0}}} tan^{-1}\left\{\frac{\left(\sqrt{KC_{D0}}\right)\zeta C_{L1}}{C_{D1} - \zeta KC_{L1}^2}\right\} \qquad (7.25)$$

The drag and lift coefficients at the beginning of the cruise are C_{D1} and C_{L1}.

<u>Example 7.6</u>

In an airplane, $W_1/S = 40\ lb/ft^2$, $C_{D0} = 0.02$, $K = 0.05$, $\eta = 0.85$, $c_P = 0.5\ lb/h.HP$. Maximum power to weight ratio is 45ft/s at sea level. Determine the fuel consumption for a cruise of 1500 miles at 75% of maximum power to weight ratio and at an altitude of 10000ft. Estimate the time of flight. ($\rho = 0.001756\ slug/ft^3$)

$$\left(\frac{P}{W}\right)_{10000} = 0.75 \times 45 \times \frac{0.001756}{0.002377} = 24.93\frac{ft}{s}$$

We follow the procedure in Example 7.2 to find the airspeed graphically.

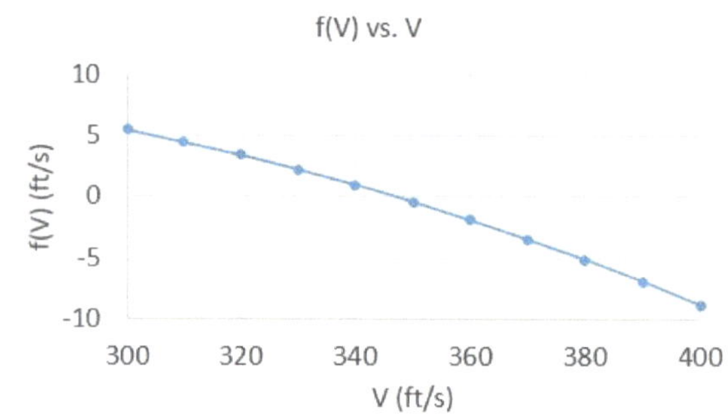

Figure 7.4a: Example 7.6

$$V = 350\ {}^{ft}/_{s} = 239\ mph$$

The time of flight is (compare with the time of flight in Example 7.5)

$$\Delta t = \frac{1500}{239} = 6.28\ hours$$

As this is a constant altitude constant airspeed flight, (lift coefficient changes during the flight) we will find the "cruise fuel weight fraction" from Eqn. (7.25). The dynamic pressure is

$$Q = 0.5 \times 0.001756 \times 350^2 = 79.02\ {}^{lb}/_{ft^2}$$

$$C_{L1} = \frac{40}{79.02} = 0.5062$$

$$C_{D1} = 0.02 + 0.05 \times 0.5062^2 = 0.03281$$

$$c_P = 0.5\frac{lb}{h.HP} \times \frac{1HP}{550\ {}^{lb.ft}/_{s}} \times \frac{1h}{3600s} \times \frac{5280ft}{1\ mile} = \frac{0.5 \times 5280}{550 \times 3600}\ per\ mile$$

We will find ζ from Eqn. (7.25) by rewriting the equation as

$$f(\zeta) = X_{V,\rho} - \frac{\eta}{c_P}\frac{1}{\sqrt{KC_{D0}}}tan^{-1}\left\{\frac{\left(\sqrt{KC_{D0}}\right)\zeta C_{L1}}{C_{D1} - \zeta K C_{L1}^2}\right\} \; miles$$

The figure below shows the plot of $f(\zeta)$ versus zeta.

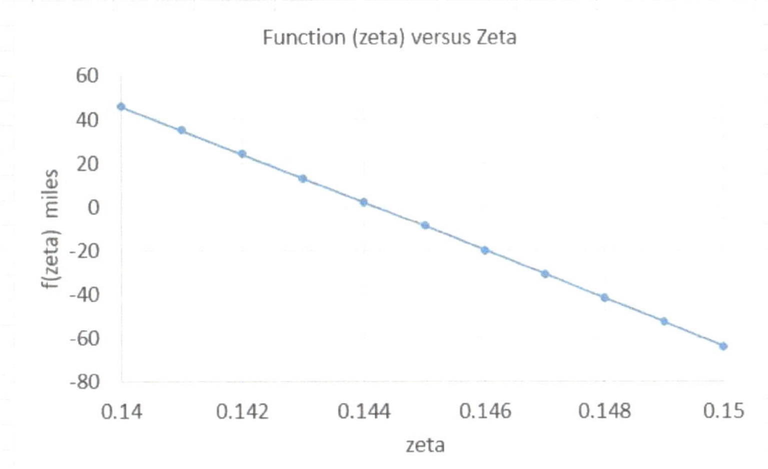

Figure 7.4b: Example 7.6

The graphical solution for $f(\zeta) = 0$ is $\zeta = 0.144$ (compare with the cruise fuel weight fraction in Example 7.5).

Observation:

- In Example 7.6, the fuel consumption is *7.3%* more than the fuel consumption in Example 7.5 and the time of flight in Example 7.6 is *27%* less than the time of flight in Example 7.5. We burn more 7.3% more fuel to shorten the flight time by 27%.

7.4 Takeoff and Climb

7.4.1 Ground Run

We can determine the ground run for a propeller airplane by making the following substitution in Eqn. (4.9)

$$\frac{T}{W_{TO}} = \frac{P}{W_{TO}}\frac{1}{V_{TO}}$$

$$d = \frac{V_{TO}^3}{2g\left\{\left(P/W_{TO}\right) - \mu V_{TO}\right\}} \tag{7.26}$$

We determine V_{TO} from Eqn. (4.10).

Example 7.7

In an airplane, the wing loading is 40lb/ft², power to weight ratio at sea level is 45ft/s, and maximum lift coefficient of 1.6. The coefficient of rolling friction is 0.03. Determine the ground run at an altitude of 5000ft. (ρ = 0.002048slug/ft³)

Power to weight ratio at 5000ft

$$\left(\frac{P}{W}\right)_{5000} = 45 \times \frac{0.002048}{0.002377} = 38.77\frac{ft}{s}$$

$$V_{stall} = \left\{\frac{2 \times 40}{0.002048 \times 1.6}\right\}^{\frac{1}{2}} = 156.3\frac{ft}{s}$$

$$V_{TO} = 1.2V_{stall} = 187.5\frac{ft}{s}$$

$$d = \frac{187.5^3}{2 \times 32.2 \times \{38.77 - 0.03 \times 187.5\}} = 3088\ ft$$

7.4.2 Steepest Climb

For fastest climb, we maximize the rate of climb by calculating the corresponding airspeed. From Eqn. (4.21),

$$\frac{d\dot{h}}{dV} = \frac{d}{dV}(V sin\gamma) = \frac{d}{dV}\left(\frac{TV - DV}{W}\right) = \frac{d}{dV}\left(\frac{P - DV}{W}\right) = 0$$

By inserting the expression for drag, we find the airspeed for fastest climb for a fixed power to weight ratio.

$$\frac{d}{dV}\left\{-VQSC_{D0} - KV\frac{W^2}{QS}\right\} = -QSC_{D0} - VSC_{D0}\frac{dQ}{dV} - K\frac{W^2}{QS} + KV\frac{W^2}{Q^2S}\frac{dQ}{dV} = 0$$

By using Eqn. (4.23)

$$-3QSC_{D0} + KV\frac{W^2}{QS} = 0$$

The dynamic pressure for fastest climb is

$$Q_{FC} = \left(\frac{K}{3C_{D0}}\right)^{\frac{1}{2}}\frac{W}{S} \tag{7.27}$$

Therefore, the airspeed for fastest climb is

$$V_{FC} = \left(\frac{K}{3C_{D0}}\right)^{1/4} \left\{\frac{2(W/S)}{\rho}\right\}^{1/2} \tag{7.28}$$

The fastest rate of climb is

$$\dot{h}_{FC} = \frac{P}{W} - \frac{D_{FC}V_{FC}}{W} \tag{7.29}$$

<u>*Example 7.8*</u>
An airplane weighs 8700lb, has wing area of 280ft², K = 0.05, C_{D0} = 0.02, power to weight ratio at sea level is 43ft/s. Determine the fastest rate of climb at full power. (ρ = 0.002377slug/ft³)

$$\frac{W}{S} = \frac{8700}{280} = 31.07\frac{lb}{ft^2}$$

$$Q_{FC} = \left(\frac{0.05}{3 \times 0.02}\right)^{\frac{1}{2}} \times 31.07 = 28.36\frac{lb}{ft^2}$$

$$V_{FC} = \left(\frac{0.05}{3 \times 0.02}\right)^{1/4} \left\{\frac{2 \times 31.07}{0.002377}\right\}^{1/2} = 154.5\frac{ft}{s}$$

$$C_{L,FC} = \frac{W}{Q_{FC}S} = 1.095$$

$$C_{D,FC} = C_{D0} + KC_{L,FC}^2 = 0.08$$

$$D_{FC} = Q_{FC}SC_{D,FC} = 635.3\ lb$$

$$\dot{h}_{FC} = \frac{P}{W} - \frac{D_{FC}V_{FC}}{W} = 31.72\frac{ft}{s}$$

7.5 Descent/Approach and Landing

7.5.1 Approach

We can calculate the landing airspeed from Eqn. (5.4), the dynamic pressure at landing from Eqn. (5.5), and the drag at landing from Eqn. (5.6). To calculate the power at landing, we modify Eqn. (5.7) as follows.

$$\frac{P_{LA}}{W_{LA}} = \left(\frac{D_{LA}}{W_{LA}} - sin\theta\right) V_{LA} \tag{7.30}$$

The approach angle is θ .

7.5.2 Ground Run

We can calculate the landing airspeed from Eqn. (5.4). To calculate the ground run, we modify Eqn. (5.14) as follows.

$$s_g = \frac{V_{LA}^3}{2g\left\{\left({P_R}/{W_{LA}}\right) + \mu V_{LA}\right\}} \tag{7.31}$$

Reversed power from a reversible propeller is about 40% of the full power. We have not included the effect of spoilers.

7.6 Turning Flight

We will do the analysis only for the tightest turn. We use the relation $P = TV$ and re-write Eqn. (6.13) as

$$n^2 = \frac{\frac{1}{2}\rho V\left(\frac{P}{W}\right)\left(\frac{W}{S}\right) - \frac{1}{4}\rho^2 V^4 C_{D0}}{K\left(\frac{W}{S}\right)^2} \tag{7.32}$$

By differentiating Eqn. (7.32) with respect to V

$$2n\frac{dn}{dV} = \frac{\frac{1}{2}\rho\left(\frac{P}{W}\right)\left(\frac{W}{S}\right) - \frac{3}{4}\rho^2 V^3 C_{D0}}{K\left(\frac{W}{S}\right)^2} \tag{7.33}$$

By inserting Eqns. (7.32) and (7.33) in the condition for tightest turn, Eqn. (6.16),

$$f(V) = \frac{1}{4}\rho^2 V_{TT}^4 C_{D0} + K\left(\frac{W}{S}\right)^2 - \frac{3}{8}\rho V_{TT}\left(\frac{P}{W}\right)\left(\frac{W}{S}\right) = 0 \tag{7.34}$$

For a given power to weight ratio, we solve Eqn. (7.34) graphically for V_{TT}. For this airspeed, we find the load factor n_{TT} from Eqn. (7.32) and the lift coefficient from

$$C_{L,TT} = \frac{2n_{TT}\left(\frac{W}{S}\right)}{\rho V_{TT}^2} \tag{7.35}$$

Valid Turn

- Load factor is more than one.
- Lift coefficient is less than the maximum lift coefficient.

We find the radius of the turn from Eqn. (6.25).

$$r_{TT} = \frac{V_{TT}^2}{g} \frac{1}{\sqrt{n_{TT}^2 - 1}} \quad (7.36)$$

<u>Example 7.9</u>

In an airplane, wing loading is 45lb/ft², K = 0.05, C_{D0} = 0.02. Determine the airspeed, load factor, lift coefficient, and radius for tightest turn at sea level for various power-weight ratios.

We will only show the results without the details of the calculations and examine the trend in the solutions when the power weight ratio is changed.

P/W	V(TT)	CL(TT)	n(TT)	r(TT)
18.2	174.9	1.474	1.191	1465
18.4	166.8	1.612	1.185	1358
18.6	161.4	1.717	1.182	1285
18.8	157.1	1.809	1.179	1226
19	153.4	1.894	1.177	1177

Figure 7.5: Example 7.9

Observations for Tightest Turn

- There is no solution to this problem for power weight ratio less than 18.2.
- The airspeed decreases when the power to weight ratio increases.
- Lift coefficient increases when power to weight ratio increases. The airplane will stall beyond a maximum power weight ratio.
- Load factor decreases when power weight ratio increases.
- Radius of tightest turn decreases when power weight ratio increases.
- The airplane will stall below a minimum radius of turn.

7.7 Engine and Propeller

Detailed propeller design, such as, number of blades, fixed pitch or variable pitch, and blade angle at *0.75R,* are beyond the scope of this text. We will simply demonstrate the process of choosing an engine and matching a three-blade propeller with the engine for our need. We will consider a numerical example to demonstrate this process for better understanding.

Requirements

- An airplane spends most of its time in cruising. Therefore, we choose a propeller that suits best to the cruise condition.
- The designer must have access to the engine performance data that the engine manufacturer supplies. We will use the engine performance data for Lycoming engines that are available on the web. The author found the URLs by entering "Lycoming aircraft engine performance data" in a Google search.

- The designers must have access to their own propeller testing data available from wind tunnel testing. We will use the propeller testing data for a three-blade propeller (Biermann and Hartman, 1937).

Consider an *8750lb* airplane with wing area of *280ft²*, *K = 0.05,* and C_{D0} = *0.02.* We first find the power required for best range at an altitude of *10000ft* ($\rho = 0.001756\ slug/ft^3$).

$$E_M = \sqrt{\frac{1}{4KC_{D0}}} = 15.8$$

$$\frac{W}{S} = \frac{8730}{280} = 31.18\frac{lb}{ft^2}$$

$$C_{L,EM} = \sqrt{\frac{0.02}{0.05}} = 0.63$$

$$V_{BR,CL} = \sqrt{\frac{2 \times 31.18}{0.001756 \times 0.63}} = 237.4\,{}^{ft}/_{s}$$

From Eqn. (7.24)

$$P_{BR} = \frac{V_{BR,CL}W}{E_M} = \frac{237.4 \times 8750}{15.8} = 131500\ {}^{lb.ft}/_{s}$$

The designers chose two Lycoming IO-540 Series engines for this airplane. We assume that the efficiency of the engine-propeller system is 80%. Therefore, the power from each engine is

$$P_E = \frac{131500}{0.8 \times 2} = 82190\ {}^{lb.ft}/_{s} = 149.4\ HP$$

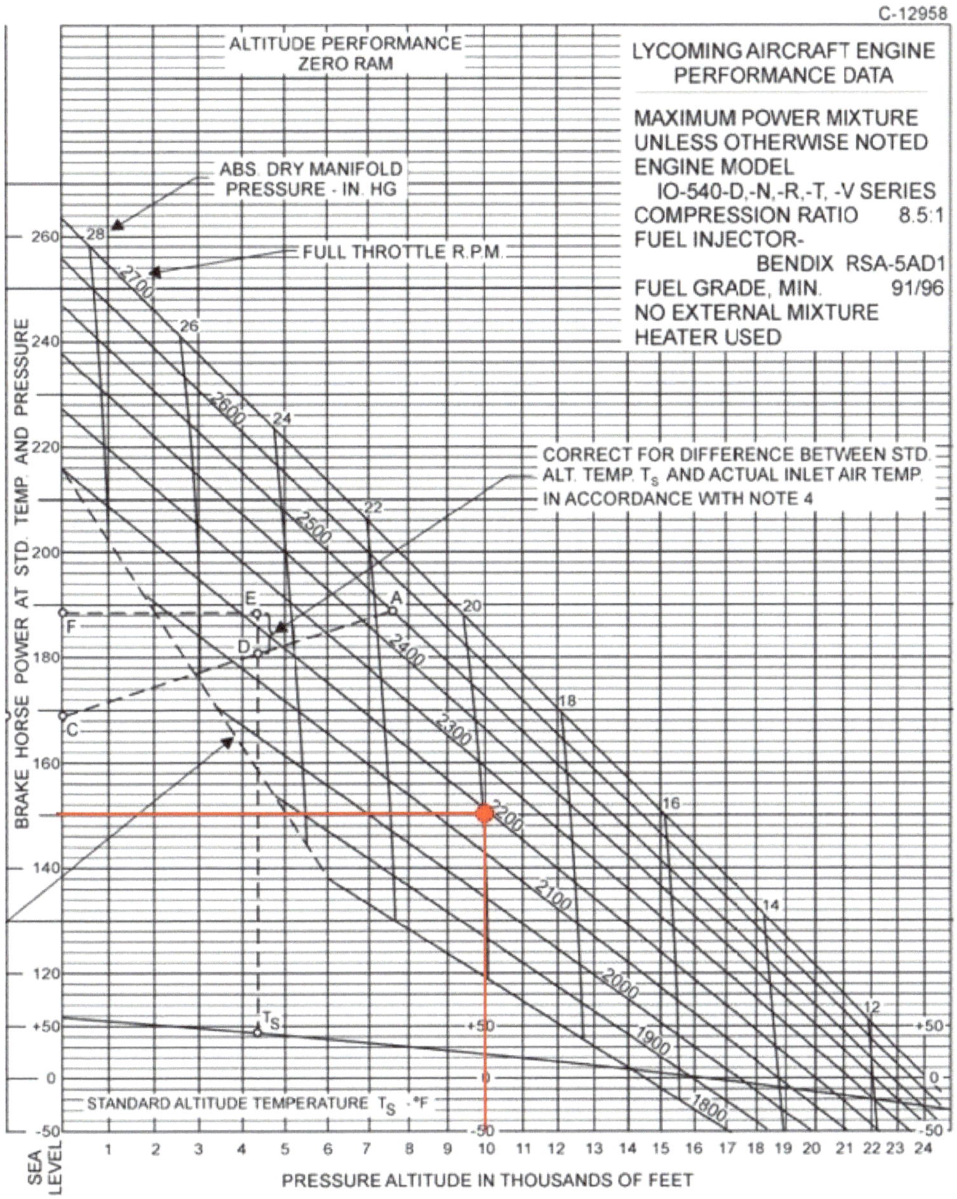

Figure 7.6: Engine performance.

On the engine performance chart from Lycoming, we locate the point with coordinates *(10000ft, 150HP)*. The RPM corresponding to this operating condition is *2200.* The revolutions per second is

$$n = \frac{2200}{60} = 36.67$$

We calculate the "speed power coefficient" as

$$C_s = \left(\frac{\rho V^5}{P_E n^2}\right)^{1/5} = \left(\frac{0.001756 \times 237.4^5}{82190 \times 36.67^2}\right)^{1/5} = 1.643$$

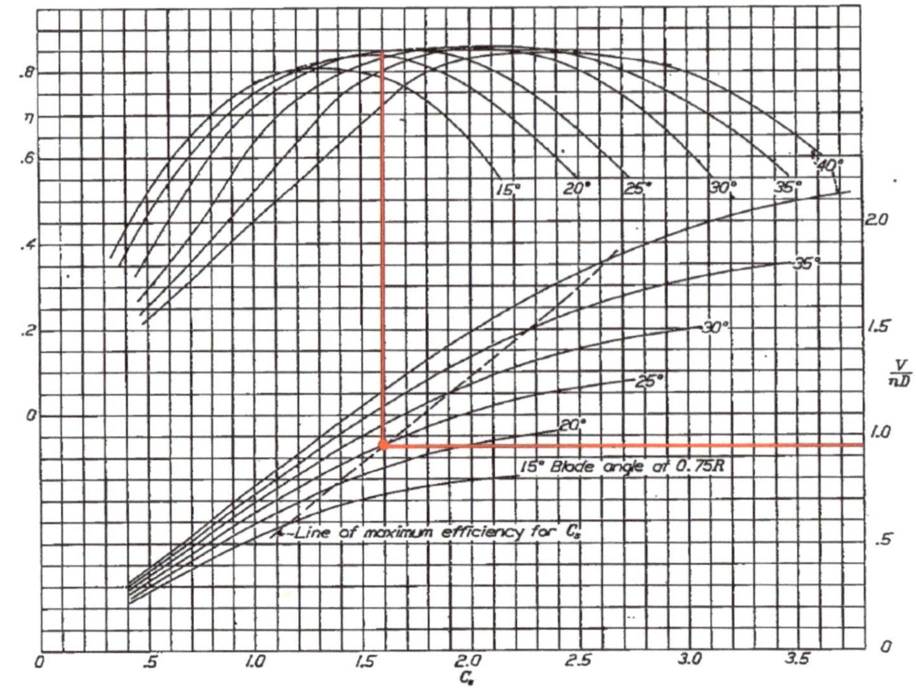

FIGURE 13.—Design chart for propeller 5868-9, 3 blades, radial engine nacelle.

Figure 7.7 Propeller performance.

On the propeller performance chart, we locate a point on the dotted line of maximum efficiency corresponding to $C_s = 1.6$. For this point, we read the value of the advance ratio from the vertical axis on the right. From the definition of advance ratio,

$$J = 0.95 = \frac{V}{nD} = \frac{237.4}{36.67 \times D}$$

We solve this equation for the first estimate of the propeller diameter as $D = 6.8ft$. Furthermore, from the propeller performance chart, the blade angle at $0.75R$ is 25^0 and efficiency is 0.85. To get an improved estimate of the diameter, the designer can repeat the calculation with the efficiency of 0.85.

To find the specific fuel consumption c_P, we use the fuel consumption chart for the engine.

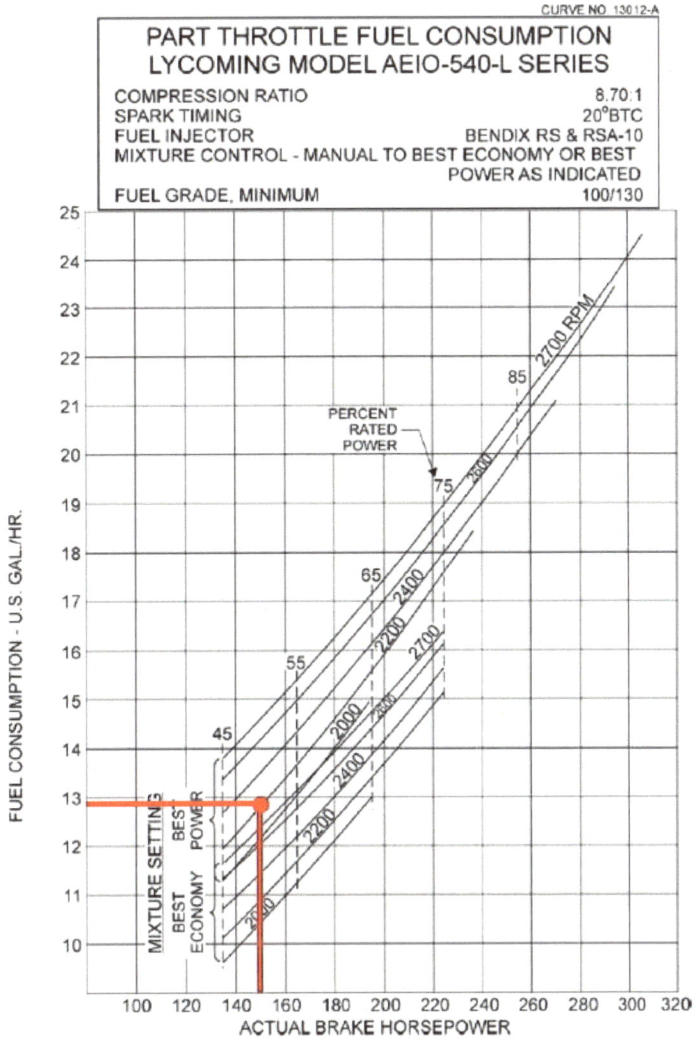

Figure 7.8: Fuel consumption.

We locate a point on the *2200-RPM* line that corresponds to *150HP* and read the fuel consumption rate from the vertical axis as *12.9gallons/h*. Specific weight of fuel is about *6.3lb/gallon*. Thus

$$c_P = \frac{12.9 \times 6.3}{P_E} = \frac{12.9 \times 6.3}{150} = 0.5418\ {}^{lb}/_{h.HP}$$

8 Jet Propulsion

8.1 Engines

An engine converts chemical energy of a fuel into mechanical energy that propels or pushes an airplane forward. Feasibility studies are under way to replace the fuel with batteries and solar cells. Airplane propulsion is of two kinds: propellers and air breathing. In this chapter, we will focus on air breathing or jet engines only. Furthermore, we will analyze an ideal cycle. This analysis includes the thermodynamics of the airflow through the engine and excludes the details of the components, such as, the compressor, the combustor, and the turbine.

We will begin with a brief review of the necessary thermodynamic relations.

8.2 Thermodynamics

8.2.1 System and State Variables

The system is a lump of air that enters the engine, undergoes various changes in the engine, and eventually exits the engine. For continuous engine operation, a series of such lumps moves through the engine.

A state variable is a macroscopic property, such as pressure, density, and temperature, of the system that depends only on the current, equilibrium state of the system. We can measure the macroscopic properties without considering the details of the molecular motion in the system. When a state variable has the same value in all the parts of the system, the system is in equilibrium.

A process is reversible when a system is in equilibrium at the entrance and at the exit of a process. During the process, the system interacts with its surroundings.

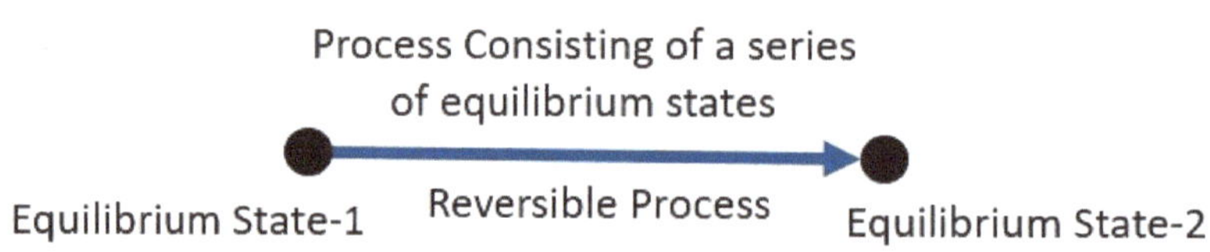

In our ideal analysis, all the processes in the engine, that the lump of air undergoes, are reversible.

8.2.2 Ideal Gas Law

The ideal gas law connects three state variables, pressure, density, and temperature. The units for these state variables are pressure in lb/ft^2, density in $slug/ft^3$, and temperature in $^oR = 459.7 + {^oF}$.

$$p = \rho RT$$

Specific gas constant for air is

$$R = 287\,\frac{N.m}{kg.K} = 1716\frac{lb.ft}{slug.R}$$

We define specific volume as volume of air for unit mass

$$v = \frac{V}{m} = \frac{1}{\rho}$$

The perfect gas law in terms of specific volume is

$$pv = RT \tag{8.1}$$

8.2.3 Internal Energy

Internal energy is the total energy of the molecules contained in the system. This includes: (i) translation kinetic energy of the molecules that are in random motion, (ii) vibration energy of polyatomic molecules, (iii) rotational energy of the molecules, and (iv) intra- or inter-molecular potential energy.

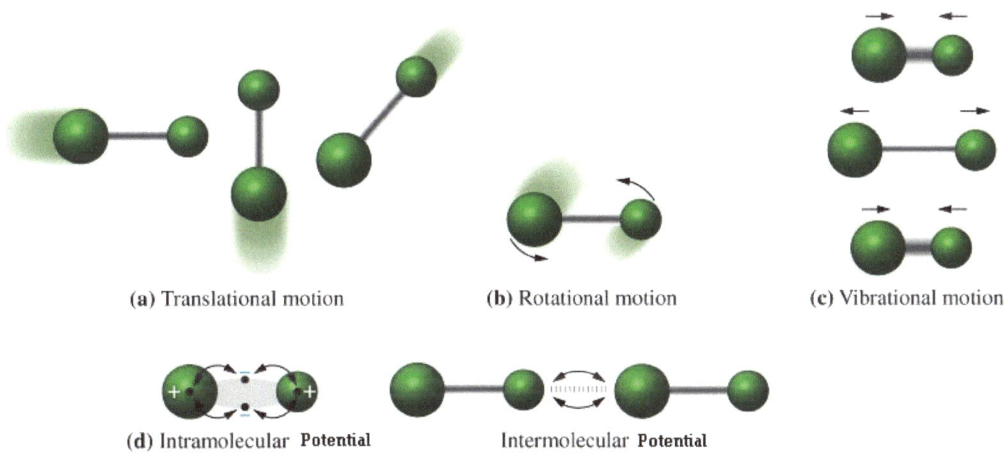

For a calorically perfect gas, internal energy is a function of temperature.

$$e = C_v T \tag{8.2}$$

The specific heat $C_v = 4290\ lb.ft/slug.R$ is a constant property of air.

8.2.4 Enthalpy

One combination of state variables appears frequently in our analysis. We give this combination a special name: enthalpy.

$$h = e + \frac{p}{\rho}$$

By using Eqns. 8.1 and 8.2

$$h = C_v T + RT = (C_v + R)T = C_p T \tag{8.3}$$

The specific heat

$$C_p = C_v + R \tag{8.4}$$

The specific heat $C_p = 6006\, lb.ft/slug.R$ is a constant property of air. We define the ratio of specific heats as

$$\gamma = \frac{C_p}{C_v} = 1.4\; for\; air \tag{8.5}$$

By combining Eqns. (8.4) and (8.5)

$$C_v = \frac{R}{\gamma - 1} \quad C_p = \frac{\gamma R}{\gamma - 1} \tag{8.6}$$

8.2.5 Equilibrium and Non-equilibrium Work

Consider a container with a piston that has a pile of sand. The piston keeps air in the container in State-1 with pressure, volume, and temperature p_1, V_1, T_1. The area of cross section of the container is A. When we remove one grain of sand, the piston rises by a distance dh. The increase in the volume of air is

$$dV = Adh > 0$$

We continue this process millions of times until air reaches State-2 with pressure, volume, and temperature p_2, V_2, T_2.

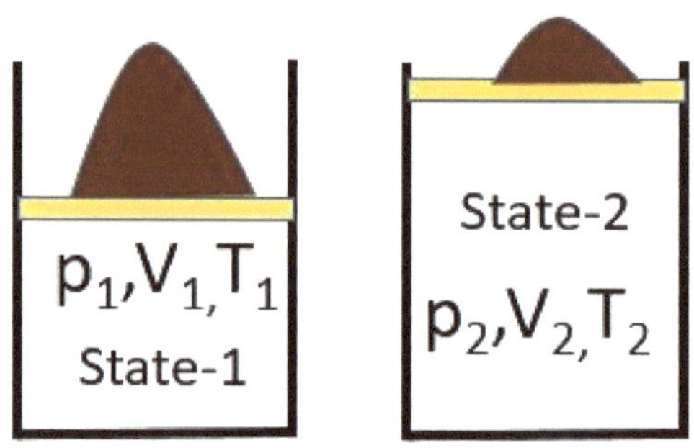

In this gradual process, air remains in equilibrium at every step of the process. At a step of the process, the pressure of air is the equilibrium pressure p and the force from air on the piston is

$$F = pA$$

Work done by air on piston is

$$\delta W = Fdh = pAdh = pdV > 0$$

The total work done by air on piston is

$$W_{air,12} = \int_1^2 \delta W = \int_1^2 pdV > 0$$

The pressure appearing under the integral is the pressure of air in equilibrium. We can reverse this process by bringing air from State-2 to State-1. At each step of this process, we add one grain of sand and the piston descends by a distance *dh*. The decrease in the volume of air is

$$dV = Adh < 0$$

After millions of steps, air reaches State-1. At each step, air remains in equilibrium and the total work done by piston on air is

$$W_{piston,21} = \int_2^1 pdV < 0$$

Observations:

- *Work done by air on piston is positive. Air loses energy.*
- *Work done by piston on air is negative. Air gains energy.*
- $|W_{air,12}| = |W_{piston,21}|$
- *Net change in energy of air is zero in the cyclic process State-1 to State-2 to State-1.*
- *The process leaves no change in either air or its surroundings.*
- *Process consists of a series of equilibrium states and is reversible.*

Consider another container with cross sectional area *A* with a piston. We can load the piston with two blocks or one block. Weight of each block is *L*.

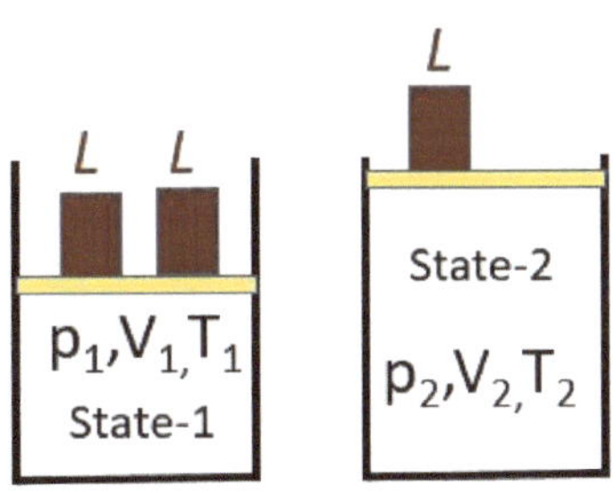

With two blocks on the piston, the pressure in air is

$$p_1 = \frac{2L}{A}$$

With one block on the piston, the pressure in air is

$$p_2 = \frac{L}{A}$$

In the process from State-1 to State-2, we abruptly remove one block. Air suddenly expands in a non-equilibrium process. With one block on the piston, air pushes the piston up with force

$$F = L = p_2 A$$

The total work done by air on piston is

$$W_{air,12} = \int_1^2 F dh = \int_1^2 p_2 A dh = p_2 \int_1^2 dV > 0$$

In the process from State-2 to State-1, we abruptly add one block. Air suddenly contracts in a non-equilibrium process. With two blocks on the piston, piston pushes the air down with force

$$F = 2L = p_1 A$$

The total work done by piston on air is

$$W_{piston,21} = \int_1^2 F dh = \int_1^2 p_1 A dh = p_1 \int_2^1 dV < 0$$

Observations:

- *Work done by air on piston is positive. Air loses energy.*
- *Work done by piston on air is negative. Air gains energy.*
- $|W_{air,12}| \neq |W_{piston,21}|$
- *Net change in energy of air is non-zero in the cyclic process State-1 to State-2 to State-1.*
- *The process leaves change in air or its surroundings.*
- *Process consists of a series of non-equilibrium states and is irreversible.*

8.2.6 First Law of Thermodynamics

The internal energy of air increases when heat is added to the air and internal energy decreases when heat is removed from the air. Furthermore, the internal energy of air decreases when the air does work and internal energy increases when work is done on air.

$$dE = \delta Q - \delta W$$

When we write the first law of thermodynamics for unit mass of air, we use lower case symbols

$$de = \delta q - \delta w \tag{8.7}$$

When heat is added, $\delta q > 0$, the internal energy of air increases and when heat is removed, $\delta q < 0$, the internal energy of air decreases. When air does work, $\delta w > 0$, the internal energy of air decreases and when work is done on the air, $\delta w < 0$, the internal energy of air increases. For unit mass of air, we write

$$\delta w = p dv$$

For reversible process, the pressure in the work equation is the equilibrium pressure of air.

8.2.7 Adiabatic Reversible Process

In an adiabatic process, there is no heat addition or subtraction. Therefore, the first law of thermodynamics becomes

$$de + pdv = 0$$

By using Eqn. (8.2),

$$C_v dT + pdv = 0 \tag{8.8}$$

By differentiating Eqn. (8.1),

$$pdv + vdp = RdT \;\; ; \;\; pdv = RdT - vdp = RdT - \frac{RT}{p}dp \tag{8.9}$$

By combining Eqns. (8.8) and (8.9)

$$C_v dT + RdT - \frac{RT}{p}dp = 0 \;\; ; \;\; C_p dT - \frac{RT}{p}dp = 0 \;\; ; \;\; C_p\frac{dT}{T} - R\frac{dp}{p} = 0 \tag{8.10}$$

By integrating Eqn. (8.10)

$$C_p lnT - Rlnp = const.; \; ln(T^{C_p}) - ln(p^R) = const. \;\; ; \; ln\left(\frac{T^{C_p}}{p^R}\right) = const.$$

By using Eqn. (8.6)

$$ln\left(\frac{T^{\frac{\gamma R}{\gamma-1}}}{p^R}\right) = ln\left(\frac{T^{\frac{\gamma}{\gamma-1}}}{p}\right)^R = const. \;\; ; \;\; T^\gamma p^{1-\gamma} = const. \tag{8.11}$$

By writing Eqn. (8.11) between two states

$$T_a^\gamma p_a^{1-\gamma} = T_b^\gamma p_b^{1-\gamma} \;\; ; \;\; \left(\frac{T_b}{T_a}\right)^\gamma = \left(\frac{p_b}{p_a}\right)^{\gamma-1} \tag{8.12}$$

An adiabatic reversible process is an isentropic process and Eqn. (8.12) is the isentropic relation that connects pressure and temperature of two states.

8.3 Energy Equation

We derived the mass conservation equation (Eqn. (2.2)) and the momentum conservation equation (Eqn. (2.4)) in Chapter 2. We will now derive the energy conservation equation.

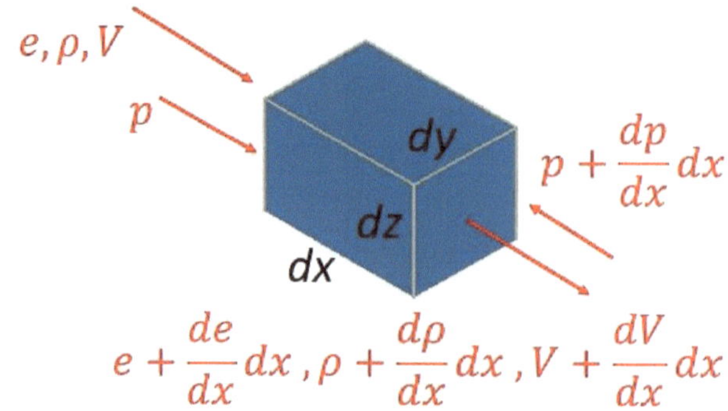

Energy consists of internal energy and kinetic energy and the energy inflow rate is

$$\rho V\left(e + \frac{1}{2}V^2\right)dydz$$

Energy outflow rate is

$$\left(\rho + \frac{d\rho}{dx}dx\right)\left(V + \frac{dV}{dx}dx\right)\left\{e + \frac{de}{dx}dx + \frac{1}{2}\left(V + \frac{dV}{dx}dx\right)^2\right\}dydz$$

By simplifying, ignoring the dx^2 and dx^3 terms, and using the mass equation (Eqn. (2.2)), energy outflow is

$$\left\{\rho V\left(e + \frac{de}{dx}dx + \frac{1}{2}V^2 + V\frac{dV}{dx}dx\right)\right\}dydz$$

Work done by pressure, in unit time, at the inlet is positive because pressure pushes air into the CV

$$(pdydz)V$$

Work done by air, in unit time, at the outlet is negative because air pushes against the pressure as it exits the CV

$$-\left\{\left(p + \frac{dp}{dx}dx\right)dydz\right\}\left(V + \frac{dV}{dx}dx\right)$$

By ignoring the dx^2 terms, the work done by air at the exit is

$$-pVdydz - V\frac{dp}{dx}dxdydz - p\frac{dV}{dx}dxdydz$$

For steady flow

$$Energy\ inflow\ rate + Pressure\ Work = Energy\ outflow\ rate$$

$$\left(-V\frac{dp}{dx} - p\frac{dV}{dx}\right)dxdydz = pV\left(\frac{de}{dx} + V\frac{dV}{dx}\right)dxdydz$$

By simplifying and using the definition of enthalpy

$$\frac{dh}{dx} + V\frac{dV}{dx} = 0$$

By integrating the energy equation between two states on the x-axis

$$h_{x1} + \frac{1}{2}V_{x1}^2 = h_{x2} + \frac{1}{2}V_{x2}^2$$

To analyze a jet engine, we include the power input from the compressor, the power output to the turbine, and the heat added by the combustor.

$$h_{x1} + \frac{1}{2}V_{x1}^2 + w_c - w_t + q = h_{x2} + \frac{1}{2}V_{x2}^2 \tag{8.13}$$

By inserting the definition of enthalpy

$$C_pT_{x1} + \frac{1}{2}V_{x1}^2 + w_c - w_t + q = C_pT_{x2} + \frac{1}{2}V_{x2}^2 \tag{8.14}$$

Important Notes

- *The unit of the terms in the energy equation, Eqn. (8.14), is ft^2/s^2*

$$\frac{ft^2}{s^2} = \frac{1}{slug}\frac{slug.ft}{s^2}ft = \frac{1}{slug}lbft = \frac{energy}{mass}$$

- *We interpret the unit of power and the rate of heat transfer as follows*

$$\frac{ft^2}{s^2} = \frac{1}{slug}\frac{slug.ft}{s^2}ft = \frac{1}{slug}lbft = \frac{s}{slug}\frac{lbft}{s} = \frac{power}{mass\ flowrate}$$

We calculate the net power of the compressor and the turbine, and the net heat added in the combustor per unit time as follows

$$\dot{W}_c = \dot{m}_a w_c\ ;\ \dot{W}_t = \dot{m}_a w_t\ ;\ \dot{Q} = \dot{m}_a q$$

8.4 Stagnation State

At the stagnation state, velocity is zero. There is no stagnation state inside the engine. We introduce this imaginary state outside the engine, for convenience in calculation.

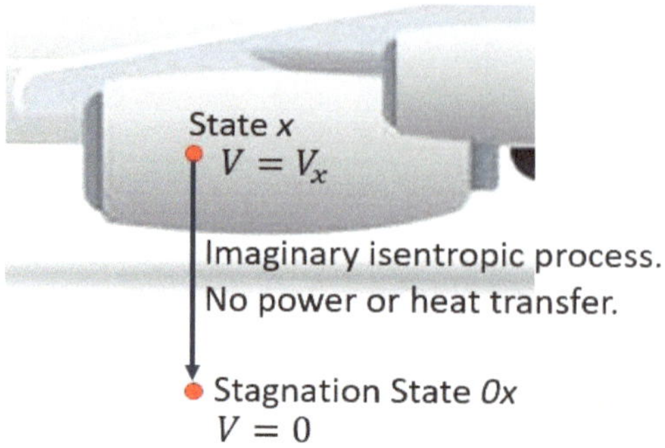

We write the energy equation between *State-x* and *State-0x*

$$C_p T_{0x} = C_p T_x + \frac{1}{2} V_x^2$$

$$\frac{T_{0x}}{T_x} = 1 + \frac{1}{2} \frac{V_x^2}{C_p T_x} = 1 + \frac{1}{2} \frac{(\gamma - 1) V_x^2}{\gamma R T_x}$$

By inserting the expression for the speed of sound and the definition of Mach number

$$a_x = \sqrt{\gamma R T_x} \;;\quad M_x = \frac{V_x}{a_x}$$

$$\frac{T_{0x}}{T_x} = 1 + \frac{(\gamma - 1)}{2} M_x^2 \qquad (8.15)$$

By using Eqn. (8.12), the isentropic relation between *State-x* and *State-0x* is

$$T_{0x}^{\gamma} p_{0x}^{1-\gamma} = T_x^{\gamma} p_x^{1-\gamma} \;;\; \frac{p_{0x}}{p_x} = \left(\frac{T_{0x}}{T_x}\right)^{\frac{\gamma}{\gamma-1}}$$

Or

$$\frac{p_{0x}}{p_x} = \left(1 + \frac{\gamma - 1}{2} M_x^2\right)^{\frac{\gamma}{\gamma-1}} \qquad (8.16)$$

8.5 Thrust

A jet engine has five components. The following description of the components is valid for an ideal jet engine.

i. Diffuser: Air enters the engine through the inlet, *State-1.* At *State-1,* the pressure is atmospheric. After entering, air passes through the diffuser that increases the pressure in air in an isentropic process.

ii. Compressor: The compressor further increases the pressure in air in an isentropic process by power addition. The pressure ratio across the compressor is a design parameter of the engine.

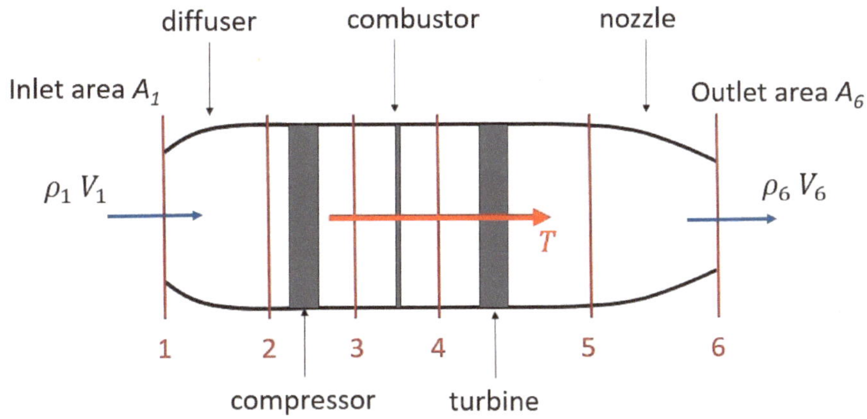

iii. Combustor: The combustor adds heat to the air by burning jet fuel in an isobaric, constant pressure, process. The temperature of air increases. The throttle setting controls the amount of fuel injection into the combustor.

iv. Turbine: The turbine reduces the pressure in air in an isentropic process by power extraction. The power extraction by the turbine is equal to the power addition by the compressor. In essence, the turbine runs the compressor. The temperature at the entrance of the turbine, *State-4*, is a design parameter of the engine. The upper limit of temperature at *State-4* limits the amount of fuel injection into the combustor.

v. Nozzle: The nozzle decreases the pressure in air so that the exit pressure, *State-6*, is equal to the atmospheric pressure, $p_6 = p_1 = \textit{atmospheric pressure}$.

The purpose of the engine is to increase the velocity of air, $V_6 > V_1$, by pushing air to the right. From Newton's Third Law, the air pushes the engine to the left. This push from air on the engine is the thrust. To derive an expression for thrust, we consider the entire engine as our CV and write the equations of conservation of mass and momentum.

Mass inflow rate and outflow rate are

$$\rho_1 V_1 A_1 \ ; \ \rho_6 V_6 A_6$$

For steady flow

$$\rho_1 V_1 A_1 = \rho_6 V_6 A_6 = \dot{m}_a \tag{8.17}$$

Example-8.1
An airplane's flight altitude is 20000 ft and flight speed is 400 ft/s. The inlet area of the engine is 10 ft². Determine the mass flowrate of air.

$$\rho_{20000} = 0.001267 \frac{slug}{ft^3}$$

We assume that the inlet velocity is equal to the flight speed.

$$\dot{m}_a = 0.001267 \times 400 \times 10 = 5.068 \frac{slug}{s}$$

The momentum inflow rate and outflow rate are

$$\dot{m}_a V_1 \; ; \; \dot{m}_a V_6$$

Force from pressure on air is zero, as the atmospheric pressure cancels out from all sides. Force from the engine on air is T. The momentum conservation equation becomes

$$\dot{m}_a V_1 + T = \dot{m}_a V_6$$

$$T = \dot{m}_a (V_6 - V_1) \tag{8.18}$$

Example-8.2
The inlet area of an engine is 10 ft², the inlet and exit velocities are 400 ft/s and 2900 ft/s. Determine the thrust at sea level and at an altitude of 20000 ft.

$$\rho_{SL} = 0.002377 \frac{slug}{ft^3} \; ; \; \rho_{20000} = 0.001267 \frac{slug}{ft^3}$$

$$\dot{m}_{a,SL} = 0.002377 \times 400 \times 10 = 9.508 \frac{slug}{s}$$

$$\dot{m}_{a,20000} = 0.001267 \times 400 \times 10 = 5.068 \frac{slug}{s}$$

$$T_{SL} = 9.508 \times (2900 - 400) = 23770 \; lb$$

$$T_{20000} = 5.068 \times (2900 - 400) = 12670 \; lb$$

Observation:

$$\frac{T_{20000}}{T_{SL}} = \frac{\rho_{20000}}{\rho_{SL}} \tag{8.19}$$

Thrust specific fuel consumption (TSFC) is the mass of fuel burnt in unit time for unit thrust.

Example-8.3
An engine produces a thrust of 24000 lb and burns fuel at a rate of 0.16 slug/s. Determine the TSFC of the engine. (Jet fuel weighs about 6.8 lbm per gallon)

$$TSFC = \frac{0.16}{24000}\frac{slug}{s.lb} = \frac{0.16 \times 3600}{24000}\frac{slug}{h.lb} = \frac{0.16 \times 3600 \times 32.2}{24000}\frac{lbm}{h.lb} = 0.7728\frac{lbm}{h.lb}$$

8.5 Engine Analysis

8.5.1 Inlet

The relevant quantities at the inlet are density ρ_1, temperature T_1, pressure p_1, velocity V_1 and inlet area A_1. The mass inflow rate is

$$\dot{m}_a = \rho_1 V_1 A_1 \tag{8.20}$$

The speed of sound and Mach number are

$$a_1 = \sqrt{\gamma R T_1} \; ; \; M_1 = \frac{V_1}{a_1} \tag{8.21}$$

We use Eqns. (8.15) and (8.16) to determine the stagnation conditions at the inlet.

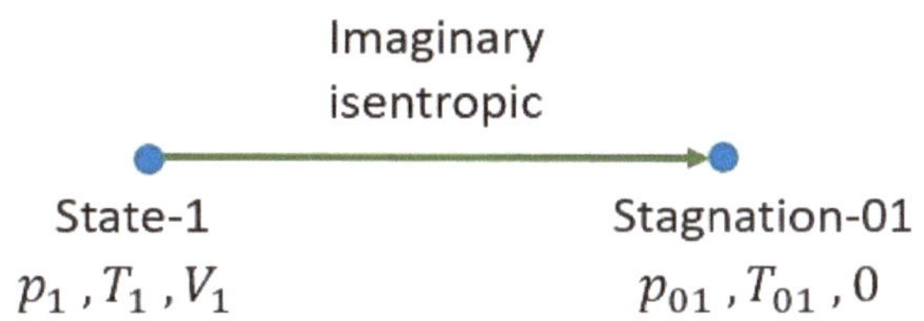

$$\frac{T_{01}}{T_1} = 1 + \frac{(\gamma - 1)}{2} M_1^2 \; ; \; \frac{p_{01}}{p_1} = \left(1 + \frac{\gamma - 1}{2} M_1^2\right)^{\frac{\gamma}{\gamma - 1}} \tag{8.22}$$

8.5.2 Diffuser

The flow through the diffuser is isentropic. In absence of power or heat addition in the diffuser, the energy equation, Eqn. (8.14), across the diffuser is

$$C_p T_1 + \frac{1}{2} V_1^2 = C_p T_2 + \frac{1}{2} V_2^2 \; ; \; C_p T_{01} = C_p T_{02} \; ; \; T_{02} = T_{01} \tag{8.23}$$

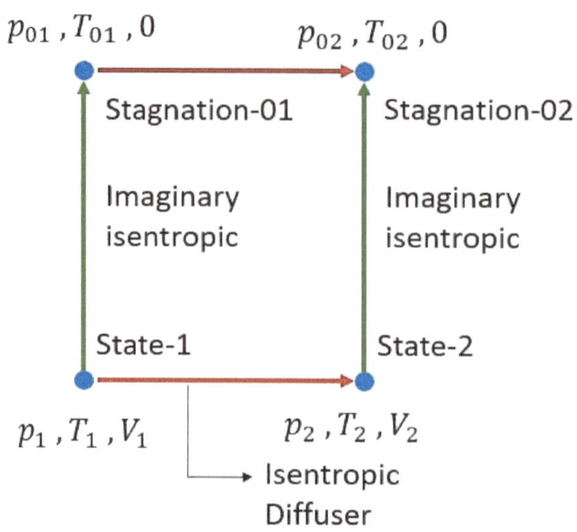

State-1 to *State-2* is isentropic, *State-1* to *State-01* is isentropic, and *State-2* to *State-02* is isentropic. Therefore, *State-01* to *State-02* is isentropic. From Eqn. (8.12),

$$\left(\frac{p_{02}}{p_{01}}\right)^{\gamma-1} = \left(\frac{T_{02}}{T_{01}}\right)^{\gamma} = 1\,;\;\; p_{02} = p_{01} \tag{8.24}$$

8.5.3 Compressor

The pressure rises across the compressor in an isentropic process. We define the pressure ratio across the compressor as

$$\frac{p_{03}}{p_{02}} = PR \tag{8.25}$$

The pressure ratio, *PR*, is a design parameter of the engine and is usually given.

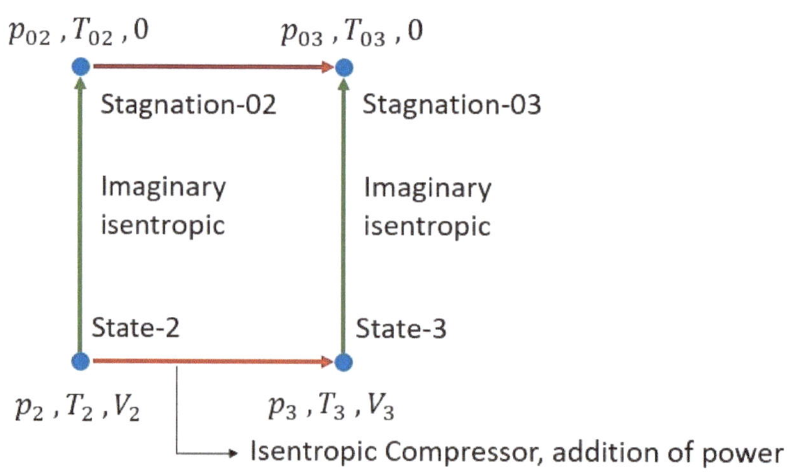

State-2 to *State-3* is isentropic, *State-2* to *State-02* is isentropic, and *State-3* to *State-03* is isentropic. Therefore, *State-02* to *State-03* is isentropic. From Eqn. (8.12),

$$\left(\frac{T_{03}}{T_{02}}\right)^{\gamma} = \left(\frac{p_{03}}{p_{02}}\right)^{\gamma-1} = PR^{\gamma-1} \; ; \; \frac{T_{03}}{T_{02}} = (PR)^{\frac{\gamma-1}{\gamma}} \tag{8.26}$$

The energy equation, Eqn. (8.14), across the compressor is

$$C_p T_2 + \frac{1}{2}V_2^2 + w_c = C_p T_3 + \frac{1}{2}V_3^2 \; ; \; C_p T_{02} + w_c = C_p T_{03}$$

Therefore,

$$w_c = C_p(T_{03} - T_{02}) \tag{8.27}$$

<u>*Example-8.4*</u>
The inlet conditions of an engine are $p_1 = 2116$ *psf,* $T_1 = 519$ *°R,* $\rho_1 = 0.002376$ *slug/ft*^{3}*,* $V_1 = 400$ *ft/s,* $A_1 = 10$ *ft*^{2}*. The pressure ratio across the compressor is 12. Determine the compressor power.*

$$\dot{m}_a = \rho_1 V_1 A_1 = \frac{9.504 \, slug}{s}$$

$$a_1 = \sqrt{1.4 \times 1716 \times 519} = 1117 \frac{ft}{s} \; ; \; M_1 = \frac{400}{1117} = 0.3582$$

From Eqn. (8.22), the stagnation conditions at the inlet are

$$p_{01} = p_1 \left(1 + \frac{\gamma - 1}{2} M_1^2\right)^{\frac{\gamma}{\gamma-1}} = 2312 \, psf$$
$$T_{01} = T_1 \left(1 + \frac{\gamma - 1}{2} M_1^2\right) = 532.3 \, °R$$

From Eqns. (8.23) and (8.24)

$$p_{02} = 2312 \, psf \; ; \; T_{02} = 532.3 \, °R$$

From Eqn. (8.25)

$$p_{03} = PR \times p_{02} = 27744 \, psf$$

From Eqn. (8.26)

$$T_{03} = T_{02} PR^{\frac{\gamma-1}{\gamma}} = 1083 \, °R$$

From Eqn. (8.27)

$$w_c = C_p(T_{03} - T_{02}) = 3.306 \times 10^6 \, lb.\frac{ft}{slug}$$

Net compressor power

$$\dot{W}_c = \dot{m}_a w_c = 31.42 \times 10^6 \; lb.\frac{ft}{s}$$

8.5.4 Combustor

The combustor burns fuel and increases the temperature of air in a constant pressure, isobaric, process. The exit temperature from the combustor is a design parameter and is given.

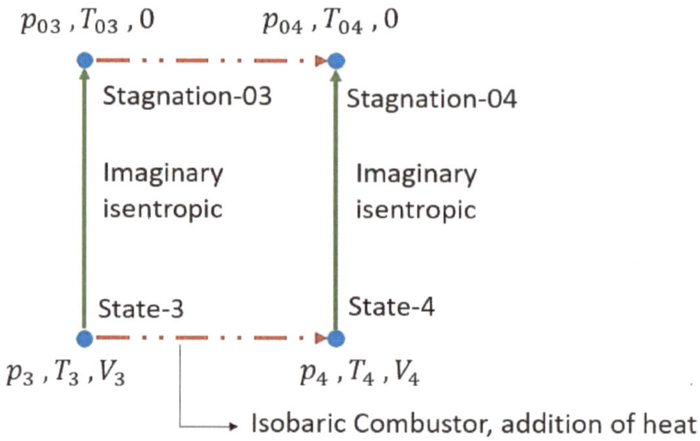

The process is isobaric, hence

$$p_{04} = p_{03} \tag{8.28}$$

The energy equation, Eqn. (8.14), across the combustor is

$$C_p T_3 + \frac{1}{2}V_3^2 + q = C_p T_4 + \frac{1}{2}V_4^2 \; ; \; C_p T_{03} + q = C_p T_{04}$$

$$q = C_p(T_{04} - T_{03}) \tag{8.29}$$

State-4 is the hottest station in the engine. We impose an upper limit on T_{04} to prevent damage in the engine. In this analysis, we assume that the engine is operating at this upper limit to maximize thrust.

The total heat input per unit time is

$$\dot{Q} = \dot{m}_a q$$

We can now relate this total heat input with the mass flow rate of fuel. We define the heating value of a fuel as the amount of heat energy that unit mass of fuel releases during combustion. Heating value of jet fuel is

$$HV = 18500 \frac{Btu}{lbm} \times \frac{778 \; lb.ft}{1 \; Btu} \times \frac{32.2 \; lbm}{1 \; slug} = 46.35 \times 10^7 \; \frac{lb.ft}{slug}$$

When fuel flow rate is $\dot{m}_f$,

$$\dot{Q} = \dot{m}_f \times HV$$

Hence,

$$\dot{m}_f = \frac{\dot{m}_a q}{HV} \tag{8.30}$$

Example-8.5 (continuation of Example 8.4)
In an engine, p_{03} = 27747 psf, T_{03} = 1083 °R, and T_{04} is limited to 2400 °R. Determine the fuel mass flow rate.

From Eqn. (8.28),

$$p_{04} = p_{03} = 27744\, psf$$

From Eqn. (8.29),

$$q = C_p(T_{04} - T_{03}) = 7.91 \times 10^6 lb.ft/slug$$

$$\dot{Q} = \dot{m}_a q = 75.19 \times 10^6 lb.\frac{ft}{s}$$

From Eqn. (8.30)

$$\dot{m}_f = \frac{\dot{Q}}{HV} = 0.1622\frac{slug}{s}$$

8.5.5 Turbine

The turbine extracts energy from the air in an isentropic process. The energy equation across the turbine is

$$h_4 + \frac{1}{2}V_4^2 - |w_t| = h_5 + \frac{1}{2}V_5^2 \;;\; C_pT_{04} - |w_t| = C_pT_{05}$$

In an ideal engine, the turbine drives the compressor. Therefore, the energy extracted by the turbine is equal to the energy added by the compressor.

$$|w_t| = w_c$$

Hence,

$$T_{05} = T_{04} - \frac{w_c}{C_p} \tag{8.31}$$

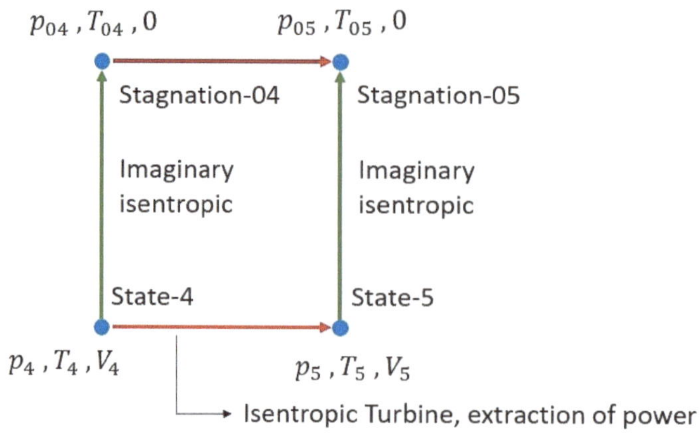

State-4 to *State-5* is isentropic, *State-4* to *State-04* is isentropic, and *State-5* to *State-05* is isentropic. Therefore, *State-04* to *State-05* is isentropic. From Eqn. (8.12),

$$\left(\frac{p_{05}}{p_{04}}\right)^{\gamma-1} = \left(\frac{T_{05}}{T_{04}}\right)^{\gamma} \tag{8.32}$$

Example-8.6 (continuation of Example 8.5)
From Example-8.4,

$$w_c = 3.306 \times 10^6\ lb.\frac{ft}{slug}$$

From Example-8.5,

$$p_{04} = 27744\ psf\ \ ;\ T_{04} = 2400\ °R$$

From Eqn. (8.31),

$$T_{05} = 2400 - \frac{3.306 \times 10^6}{6006} = 1850\ °R$$

From Eqn. (8.32),

$$p_{05} = 11150\ psf$$

8.5.6 Nozzle

The flow through the nozzle is isentropic. In absence of power or heat addition in the diffuser, the energy equation, Eqn. (8.14), across the diffuser is

$$C_pT_5 + \frac{1}{2}V_5^2 = C_pT_6 + \frac{1}{2}V_6^2\ ;\ C_pT_{05} = C_pT_{06}\ ;\ T_{06} = T_{05} \tag{8.33}$$

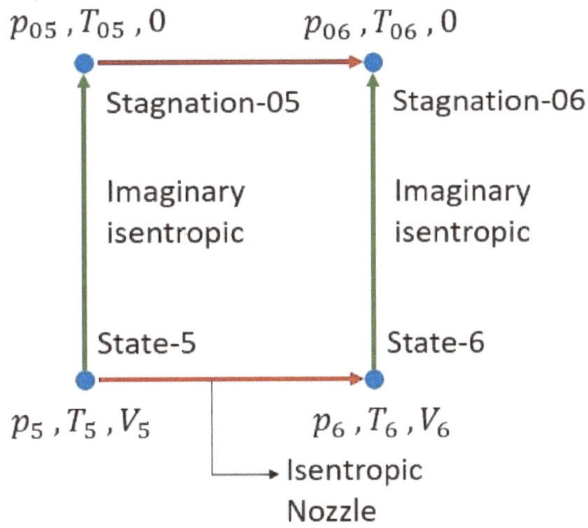

State-5 to *State-6* is isentropic, *State-5* to *State-05* is isentropic, and *State-6* to *State-06* is isentropic. Therefore, *State-05* to *State-06* is isentropic. From Eqn. (8.12),

$$\left(\frac{p_{06}}{p_{05}}\right)^{\gamma-1} = \left(\frac{T_{06}}{T_{05}}\right)^{\gamma} = 1\,;\; p_{06} = p_{05} \tag{8.34}$$

8.5.7 Exit

At the design condition that occurs at cruise speed and altitude of an ideal engine, the exit pressure is the same as the inlet pressure.

$$p_6 = p_1 = p_{atm} \tag{8.35}$$

From Eqn. (8.16),

$$1 + \frac{\gamma - 1}{2} M_6^2 = \left(\frac{p_{06}}{p_6}\right)^{\frac{\gamma-1}{\gamma}} \tag{8.36}$$

From Eqn. (8.15),

$$\frac{T_{06}}{T_6} = 1 + \frac{(\gamma - 1)}{2} M_6^2 \tag{8.37}$$

From the definitions of speed of sound and Mach number,

$$a_6 = \sqrt{\gamma R T_6}\;\;;\;\; V_6 = M_6 a_6 \tag{8.38}$$

Example-8.7 (continuation of Example 8.6)

The inlet and exit pressures in an engine are 2116 psf. The stagnation pressure and temperature at the turbine exit are p_{05} = 11150 psf and T_{05} = 1850 °R. Determine the engine thrust.

From Eqns. (8.33) and (8.34),

$$T_{06} = 1850\ °R\ ;\ p_{06} = 11150\ psf$$

From Eqn. (8.36),

$$1 + \frac{\gamma - 1}{2} M_6^2 = \left(\frac{p_{06}}{p_6}\right)^{\frac{\gamma-1}{\gamma}} = \left(\frac{11150}{2116}\right)^{\frac{\gamma-1}{\gamma}}\ ;\ M_6 = 1.743$$

From Eqn. (8.37)

$$\frac{1850}{T_6} = 1 + \frac{\gamma - 1}{2} 1.743^2\ ;\ T_6 = 1150\ °R$$

From Eqn. (8.38)

$$a_6 = \sqrt{\gamma R \times 1150} = 16662 \frac{ft}{s}\ ;\ V_6 = M_6 a_6 = 2898 \frac{ft}{s}$$

From Example-8.4,

$$\dot{m}_a = 9.504 \frac{slug}{s}$$

From Eqn. (8.18),

$$T = 9.504 \times (2898 - 400) = 23740\ lb$$

8.6 Net Energy Balance

To examine the net energy balance, we write the energy equation (Eqn. (8.14)) between *State-1* and *State-6*.

$$C_p T_1 + \frac{1}{2} V_1^2 + w_c - w_t + q = C_p T_6 + \frac{1}{2} V_6^2$$

In an ideal engine, the energy extracted by the turbine is equal to the energy added by the compressor. Therefore,

$$q = C_p (T_6 - T_1) + \frac{1}{2} (V_6^2 - V_1^2) \tag{8.39}$$

By substituting the solution from Example-8.7,

$$q = 6006 \times (1150 - 519) + 0.5 \times (2898^2 - 400^2) = 7.91 \times 10^6\ lb.\frac{ft}{slug}$$

This quantity of heat addition through combustion is equal to the quantity we calculated in Example-8.5. In a turbojet engine, all the air passes through the combustor and the combustor adds heat to the entire mass flow. In a turbofan engine, only a part of the air passes through the combustor. Rest of the air bypasses the combustor through a cowl where a fan accelerates the air.

$$\dot{m}_a = \dot{m}_{fan} + \dot{m}_{comb} \; ; \; \dot{Q} = \dot{m}_{comb} q = \dot{m}_f \times HV$$

The turbine drives the fan. Thus, in a turbofan engine, the combustor-turbine combination and the fan contribute to thrust.

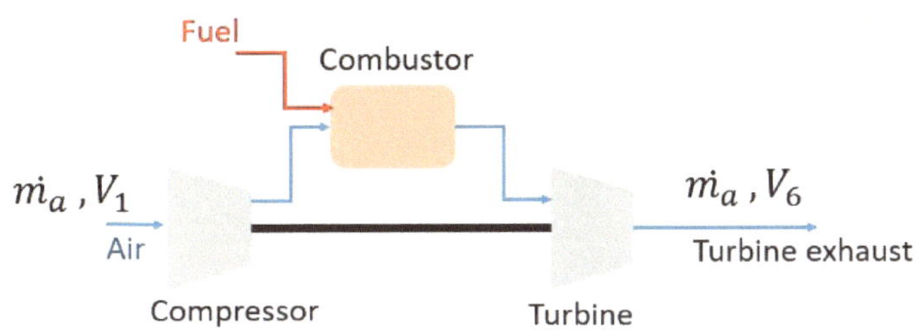

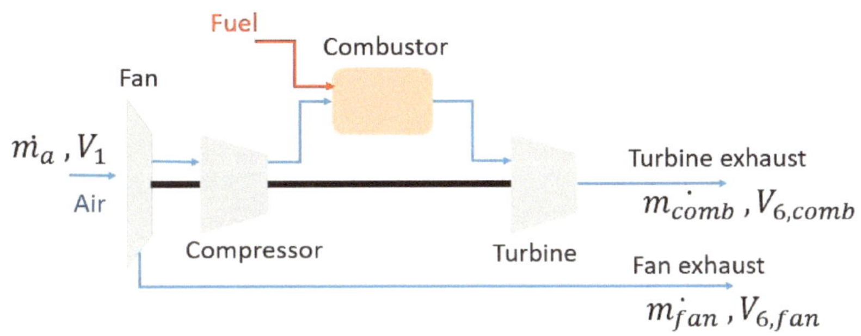

Thrust is

$$T = \dot{m}_{fan} V_{6,fan} + \dot{m}_{comb} V_{6,comb} - \dot{m}_a V_1$$

Compared to a turbojet engine, a turbofan engine generates more thrust for nearly the same amount of fuel used in the combustor.

8.6 Optimum Pressure Ratio

For given inlet conditions and combustor exit temperature (same as the turbine inlet temperature), it is possible to find a pressure ratio (*PR*) across the compressor that maximizes the thrust. We show the results of two sets of calculations. For both the sets, the inlet pressure and temperature are *2116 psf* and *519 °R* that corresponds to the sea level operation of the engine. The inlet area of the engine is *10 ft*^{2}.

i. The inlet Mach is *0.36* and the turbine inlet temperature is *2400 °R*. The thrust is maximum when $PR \approx 13$.

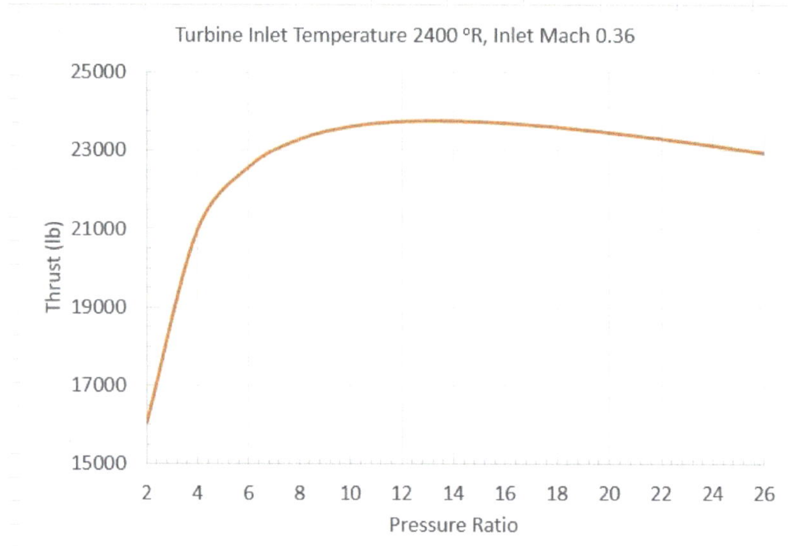

ii. The inlet Mach is *0.54* and the turbine inlet temperature is *2400 °R*. The thrust is maximum when $PR \approx 12$.

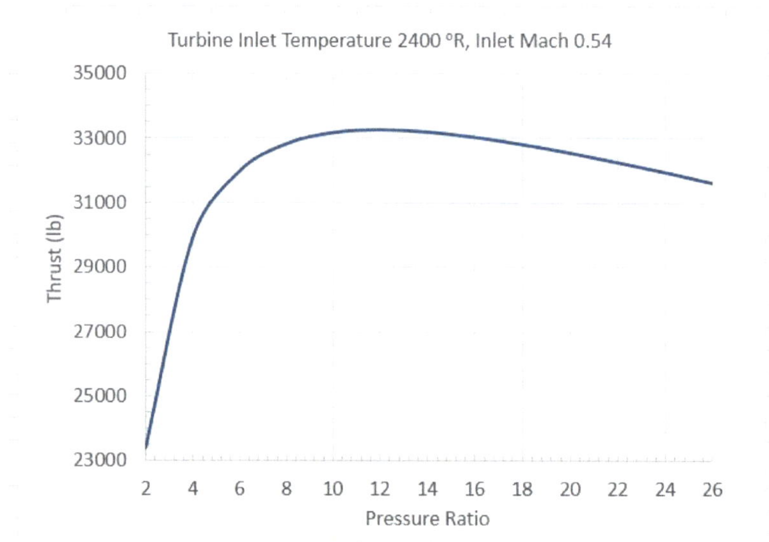

8.6 Variation of Thrust with Inlet Velocity

The following plot shows the variation of thrust with inlet velocity. The engine is operating at sea level pressure and temperature, turbine inlet temperature is *2400 °R*, pressure ratio $PR = 13$, and engine inlet area is $10\ ft^2$.

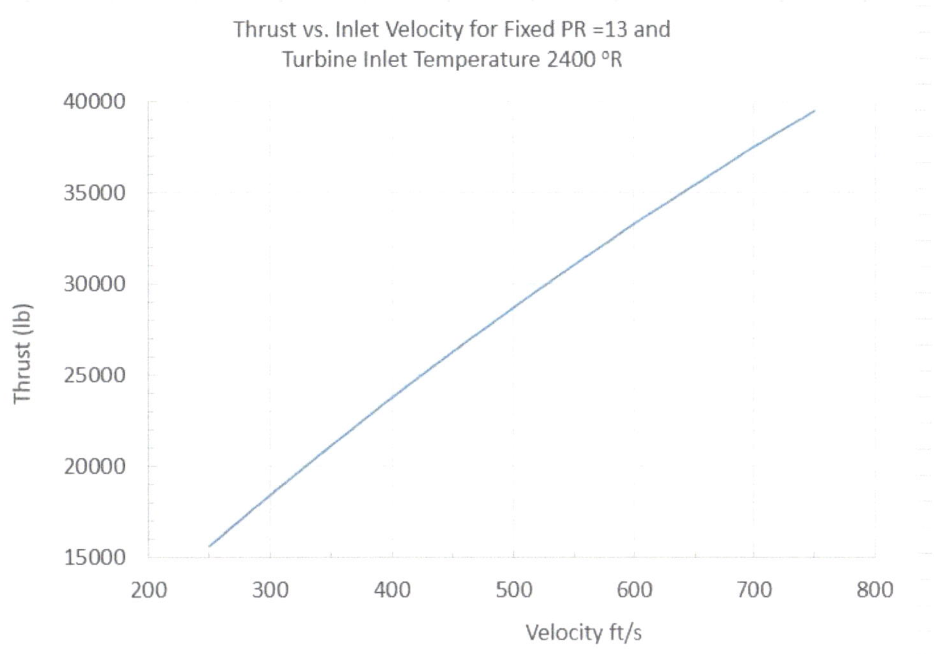

8.6 Efficiency of the Engine

We define four powers: (i) propulsive power P_P, (ii) fuel power $\dot{Q}$, (iii) power in excess kinetic energy in the wake P_{KE} , and (iv) power in excess enthalpy in the wake P_H.

$$P_P = TV_1 \quad ; \quad \dot{Q} = \dot{m}_a q$$

$$P_{KE} = \dot{m}_a \left(\frac{V_6^2}{2} - \frac{V_1^2}{2} \right) \quad ; \quad P_H = \dot{m}_a C_p (T_6 - T_1)$$

We define overall efficiency, propulsive efficiency, and thermal efficiency as

$$\eta_{overall} = \frac{P_P}{\dot{Q}} \quad ; \quad \eta_{propulsive} = \frac{P_P}{P_{KE}} \quad ; \quad \eta_{thermal} = \frac{P_{KE}}{\dot{Q}}$$

The relationship among the efficiencies is

$$\eta_{overall} = \eta_{propulsive} \times \eta_{thermal}$$

Example-8.8

An engine is operating at sea level with $p_1 = 2116$ psf, $T_1 = 519\ ^oR$. The turbine inlet temperature is $2400\ ^oR$ and pressure ratio PR = 13. Inlet area of the engine is 10 ft^2 and the inlet velocity is 400 ft/s. Determine the overall engine efficiency.

From Example-8.7 and Example-8.5,

$$T = 23740\ lb\ ;\quad \dot{Q} = 75.19 \times 10^6\ lb.\frac{ft}{s}$$

$$\eta_{overall} = \frac{23740 \times 400}{75.19 \times 10^6} = 0.1263$$

8.7 Spreadsheet

Engine analysis is long and prone to errors in manual calculation. Doing the calculations by using the Excel spreadsheet is convenient and error free.

1	p1	1456	Inlet Area	10
2	T1	483	Gamma	1.4
3	rho1	=B1/(D6*B2)	Cp	6006
4	a1	=SQRT(D2*D6*B2)	Turbine Inlet Temp	1600
5	V1	250	HV	18500
6	M1	=B5/B4	Gas Constant	1716
7	p01	=B1*(1+((D2-1)/2)*B6*B6)^(D2/(D2-1))	HP to lb.ft/s	778
8	T01	=B2*(1+((D2-1)/2)*B6*B6)		
9	mair-dot	=B3*B5*D1		
10	PR	12		
11	p02	=B7		
12	p03	=B10*B11		
13	T02	=B8		
14	T03	=B13*B10^((D2-1)/D2)		
15	wc	=D3*(B14-B13)		
16	p04	=B12		
17	T04	=D4		
18	q	=D3*(B17-B14)		
19	Q dot	=B18*B9		
20	Fuel Rate	=B19/(D5*D7*32.2)		
21	T05	=B17-B15/D3		
22	p05	=B16*(B21/B17)^(D2/(D2-1))		
23	T06	=B21		
24	p06	=B22		
25	M6-square	=((B24/B1)^((D2-1)/D2)-1)*(2/(D2-1))		
26	M6	=SQRT(B25)		
27	T6	=B23/(1+((D2-1)/2)*B25)		
28	a6	=SQRT(D2*D6*B27)		
29	V6	=B28*B26		
30	Thrust	=B9*(B29-B5)		
31	TSFC	=B20*60*60*32.2/B30		
32	Efficiency	=B30*B5/B19		

The input variables are in the white cells and the formulas for the calculations are in the yellow cells.

9 Static Stability

9.1 Basic Idea

Consider a pendulum in static equilibrium. When we displace the pendulum away from its equilibrium position, either a clockwise or a counter-clockwise restoring moment at the hinge brings the pendulum back to its equilibrium position. The pendulum has static stability.

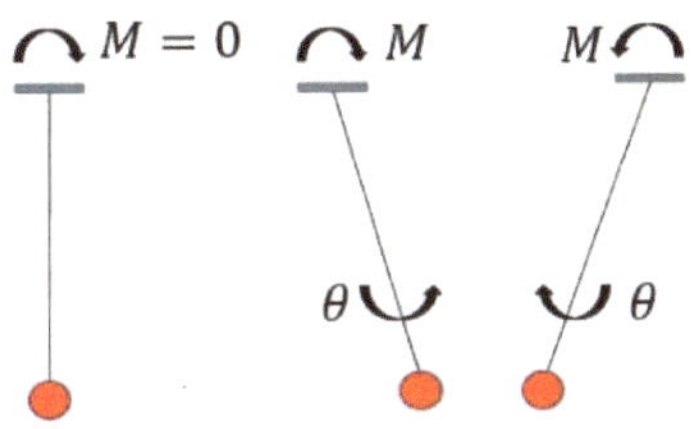

Figure 9.1: Static stability of a pendulum

When the angular displacement is counter-clockwise, the restoring moment is clockwise and when angular displacement is clockwise, the restoring moment is counter-clockwise. Therefore, the conditions of static stability are

$$M = 0 \; at \; equilibrium$$

$$\frac{dM}{d\theta} < 0 \; when \; \theta \neq equilibrium$$

Similarly, for an airplane

$$M = 0 \; at \; equilibrium$$

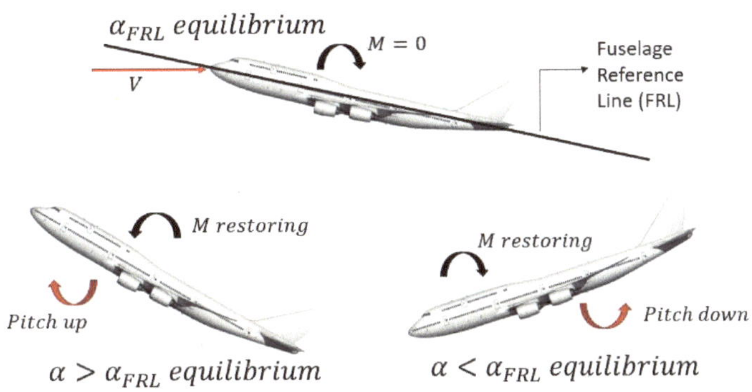

Figure 9.2: Static stability of an airplane

For static stability, when pitch is clockwise, restoring moment is counter-clockwise and when pitch is counter-clockwise, the restoring moment is clockwise.

$$\frac{dM}{d\alpha} < 0 \; when \; \alpha \neq equilibrium$$

9.2 Moment

In this analysis, clockwise moment is positive and counter clockwise moment is negative. To calculate the net moment at the center of gravity (CG) of the airplane, we will use a few approximations for simplicity.

1. *The aerodynamic centers of the wing and the tail, and the CG of the airplane are on the fuselage reference line (FRL).*
2. *We will ignore the lift and moment due to the fuselage.*
3. *Strictly speaking, the lift from the wing is perpendicular to the wing airspeed. For a small angle of attack of FRL, we assume that the lift from the wing is perpendicular to FRL.*
4. *Strictly speaking, the lift from the tail is perpendicular to the tail airspeed. For a small angle of attack of FRL, we assume that the lift from the tail is perpendicular to FRL.*
5. *The drag of the wing and the tail are along FRL. Furthermore, drag is much smaller than lift. Therefore, we do not include the drag of the wing and the tail in the moment calculations.*
6. *The pitching moment at the aerodynamic center of the tail is negligible compared to the moment due to the lift of the tail and is ignored.*

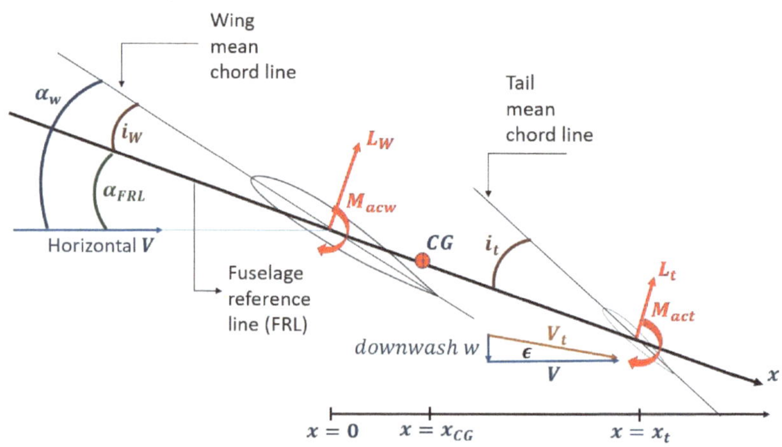

Figure 9.3: Sketch showing important variables.

Important Variables

- *The x-axis has its origin at the aerodynamic center of the wing and the axis points toward the tail. Location of the CG and the tail aerodynamic center are* x_{CG} , x_t
- *Wing airspeed V is horizontal*
- *Tail airspeed* V_t *is the vector sum of V and downwash w*
- *Downwash angle is* ϵ
- *Angle of attack of fuselage reference line is* α_{FRL}
- *Angles of incidence of wing and tail are* i_w , i_t
- *Angle of attack of wing is* $\alpha_w = \alpha_{FRL} + i_w$
- *Angle of attack of tail is* $\alpha_t = \alpha_{FRL} + i_t - \epsilon = \alpha_w - i_w + i_t - \epsilon$

- *Lift from the wing and the tail are* L_w *,* L_t
- *We measure the lift curve slopes* a_w *,* a_t *of the wing and the tail in 1/rad.*
- *Moment of the wing at the aerodynamic center is* M_{acw}

We can arrange the wing aerodynamic center, the tail aerodynamic center, and the center of gravity in two different ways on the FRL.

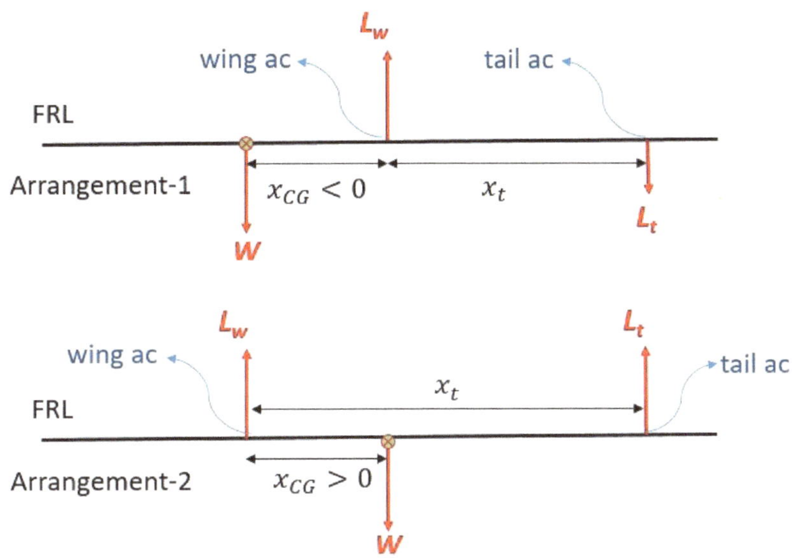

Figure 9.4: Wing tail arrangements

Observations:

- *In Arrangement-1,* $x_{CG} < 0$ *and in Arrangement-2,* $x_{CG} > 0$.
- *In Arrangement-1, the lift from the wing produces a counter clockwise (negative) moment at the CG. In Arrangement-2, the lift from the wing produces a clockwise (positive) moment at the CG.*
- *In Arrangement-1, the lift from the tail produces a clockwise (positive) moment at the CG. In Arrangement-2, the lift from the tail produces a counter clockwise (negative) moment at the CG.*
- *In Arrangement-1, the tail lift coefficient is negative. In Arrangement-2, the tail lift coefficient is positive.*

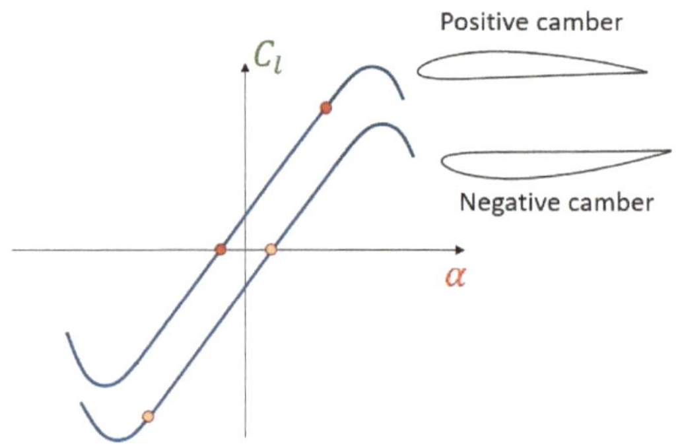

Figure 9.5: Zero lift angles for positive and negative camber

An airfoil with negative camber is suitable for Arrangement-1, as it produces negative lift for a large range of angle of attack. An airfoil with positive camber is suitable for Arrangement-2, as it produces positive lift for a large range of angle of attack.

Observations:

- *In Arrangement-1, $L_w = W + L_t$. In Arrangement-2, $L_w = W - L_t$. Therefore, lift from the wing is larger in Arrangement-1 than in Arrangement-2.*
- *For the same airspeed, the induced drag (hence thrust) is larger in Arrangement-1 than in Arrangement-2.*

9.2.1 Wing Moment

Clockwise moment at the CG from the wing is

$$\circlearrowright M_w = M_{acw} + L_w x_{CG} \tag{9.1}$$

In Arrangement-1, the second term in Eqn. (9.1) is negative, while in Arrangement-2, the second term in Eqn. (9.1) is positive. For wings with positively cambered airfoils, the pitching moment is "nose down" or counterclockwise. Hence

$$M_{acw} < 0$$

We divide Eqn. (9.1) by

$$\frac{1}{2}\rho V^2 S_w c$$

The wing area is S_w and c is the average chord of the wing.

$$\circlearrowright C_{Mw} = C_{Macw} + C_{Lw}\frac{x_{CG}}{c} = C_{Macw} + C_{Lw}h_{CG} \tag{9.2}$$

We write the lift coefficient of the wing as the product of the lift-curve slope of the wing and the wing angle of attack

$$C_{Lw} = a_w(\alpha_w - \alpha_{w0}) \tag{9.3}$$

The zero-lift angle of attack of the wing is α_{w0}. This angle is usually negative and it changes when the pilot uses the flap (see Section 2.23).

$$\alpha_w = \alpha_{FRL} + i_w \ ; \ \alpha_{FRL} = \alpha_w - i_w \tag{9.4}$$

By combining Eqns. (9.2) and (9.3)

$$\circlearrowleft C_{Mw} = C_{Macw} + a_w(\alpha_w - \alpha_{w0})h_{CG} \tag{9.5}$$

9.2.2 Tail Moment

The moment at the CG from the tail is

$$\circlearrowleft M_t = -L_t(x_t - x_{CG}) = -\frac{1}{2}\rho V_t^2 S_t C_{Lt}(x_t - x_{CG}) \tag{9.6}$$

We divide this equation by

$$\frac{1}{2}\rho V^2 S_w c$$

To get

$$\circlearrowleft C_{Mt} = -\frac{\frac{1}{2}\rho V_t^2 S_t C_{Lt}}{\frac{1}{2}\rho V^2 S_w c}(x_t - x_{CG}) = -\frac{V_t^2}{V^2}\frac{S_t}{S_w}\left(\frac{x_t}{c} - \frac{x_{CG}}{c}\right)C_{Lt}$$

Typical value of the ratio of dynamic pressures at the wing and the tail is (Raymer, 2006, pg. 472)

$$\frac{\frac{1}{2}\rho V_t^2}{\frac{1}{2}\rho V^2} = \frac{V_t^2}{V^2} = \eta = 0.9$$

We also define a "tail volume coefficient"

$$V_H = \frac{S_t}{S_w}\left(\frac{x_t}{c} - \frac{x_{CG}}{c}\right) \tag{9.7}$$

Typical tail volume coefficients for various airplanes are (Raymer, 2006, pg. 122):

Type	V_H
General Aviation	0.8
Agricultural	0.5
Cargo	1.0
Jet Transport	1.0

Figure 9.6: Tail volume coefficient

$$\circlearrowright C_{Mt} = -\eta V_H C_{Lt} \tag{9.8}$$

The angle between the wing airspeed and the tail airspeed is ϵ, the downwash angle (see equation 2.31). The angle of attack of the tail is

$$\alpha_t = \alpha_{FRL} + i_t - \epsilon = \alpha_w - i_w + i_t - \epsilon$$

We write the lift coefficient of the tail as the product of the lift-curve slope of the tail and the tail angle of attack

$$C_{Lt} = a_t(\alpha_w - i_w + i_t - \epsilon - \alpha_{t0}) \tag{9.9}$$

The zero-lift angle of attack of the tail is α_{t0} . By using Eqn. (2.26)

$$C_{Lt} = a_t\{\alpha_w - i_w + i_t - KC_{Lw} - \alpha_{t0}\} \tag{9.10}$$

By combining Eqns. (9.3), (9.8) and (9.10)

$$\circlearrowright C_{Mt} = -\eta a_t V_H[\alpha_w - i_w + i_t - Ka_w(\alpha_w - \alpha_{w0}) - \alpha_{t0}] \tag{9.11}$$

The zero-lift angle of attack α_{t0} changes when the pilot uses the elevator (see Section 9.4.1).

9.2.3 Total Moment

We write the net moment at the CG as

$$\circlearrowright C_M = C_{Mw} + C_{Mt} \tag{9.12}$$

We insert Eqns. (9.5) and (9.11) in Eqn. (9.12) and collect the terms into two groups - terms dependent on the wing angle of attack and the terms independent of the wing angle of attack.

$$\circlearrowright C_M = C_{M0} + \alpha_w C_{M\alpha} \tag{9.13}$$

Where

$$C_{M0} = C_{Macw} - a_w\alpha_{w0}h_{CG} + \eta V_H a_t(i_w - i_t - Ka_w\alpha_{w0} + \alpha_{t0}) \tag{9.14}$$

$$C_{M\alpha} = a_w h_{CG} - \eta V_H a_t(1 - Ka_w) \tag{9.15}$$

For equilibrium,

$$\circlearrowleft C_M = C_{M0} + \alpha_{w,eq} C_{M\alpha} = 0 \tag{9.16}$$

<u>*Example-9.1*</u>
In an airplane, $a_w = 5.7$, $a_t = 5.6$, $\alpha_{w0} = -0.05$ rad, $\alpha_{t0} = 0^o$, $h_{CG} = 0.2$, $V_H = 0.5$, $C_{Macw} = -0.02$. The angles of incidence of the wing and the tail are equal, and the wing aspect ratio is 7.5. Determine the wing angle of attack at equilibrium.

The span efficiency factor of the wing is

$$e = 1.78(1 - 0.045AR^{0.68}) - 0.64 = 0.8247$$

$$K = \frac{1}{\pi e AR} = 0.0515$$

From Eqn. (9.14)

$$C_{M0} = 0.07396$$

From Eqn. (9.15)

$$C_{M\alpha} = -0.6408$$

From the equilibrium Eqn. (9.16)

$$\alpha_{w,eq} = 0.1154\, rad = 6.612^o$$

<u>*Example-9.2*</u>
In an airplane, $a_w = 5.8$, $\alpha_{w0} = -0.05$rad, $i_w = 0.03$rad, $a_t = 4.8$, $\alpha_{t0} = -0.03$rad, $i_t = 0.02$, $h_{CG} = 0.2$, and $C_{Macw} = -0.02$. The wing aspect ratio is 7.5. Determine the tail volume ratio such that $\alpha_{FRL} = 0^o$.

The span efficiency factor of the wing is

$$e = 1.78(1 - 0.045AR^{0.68}) - 0.64 = 0.8247$$

$$K = \frac{1}{\pi e AR} = 0.0515$$

$$\alpha_{w,eq} = \alpha_{FRL} + i_w = 0.03$$

From Eqn. (9.5)

$$\circlearrowleft C_{Mw} = -0.02 + 5.8 \times (0.03 + 0.05) \times 0.2 = 0.0728$$

From Eqn. (9.11)

$$\circlearrowleft C_{Mt} = -0.1129 V_H$$

From the equilibrium,

$$-0.1129V_H + 0.0728 = 0\ ;\ V_H = 0.6448$$

Example-9.3

In an airplane, $a_w = 5.8$, $\alpha_{w0} = -0.05rad$, $i_w = 0.03rad$, $a_t = 4.8$, $\alpha_{t0} = -0.03rad$, $V_H = 0.7$, $h_{CG} = 0.2$, and $C_{Macw} = -0.02$. The wing aspect ratio is 7.5. Determine the tail incidence angle such that $\alpha_{FRL} = 0^o$.

The span efficiency factor of the wing is

$$e = 1.78(1 - 0.045AR^{0.68}) - 0.64 = 0.8247$$

$$K = \frac{1}{\pi eAR} = 0.0515$$

$$\alpha_{w,eq} = \alpha_{FRL} + i_w = 0.03$$

From Eqn. (9.5)

$$C_{Mw} = 0.0728$$

From equilibrium and Eqn. (9.8)

$$C_{Mt} = -0.9 \times 0.7 \times C_{Lt} = -0.0728\ ;\ C_{Lt} = 0.1156$$

From Eqn. (9.9)

$$i_t = 0.01795rad = 1.029^o$$

9.3 Static Stability

9.3.1 Neutral Point

We define the location of the neutral point of the airplane as

$$h_{NP} = \eta V_H(1 - Ka_w)\frac{a_t}{a_w} \tag{9.17}$$

With this definition and Eqn. (9.15),

$$C_{M\alpha} = a_w(h_{CG} - h_{NP})$$

$$\circlearrowright C_M = C_{M0} + \alpha_w a_w(h_{CG} - h_{NP}) \tag{9.18}$$

Observations:

- *When the center of gravity of the airplane and the neutral point of the airplane are at the same location,* $h_{CG} = h_{NP}$ *, the net moment on the airplane does not depend on the angle of attack. In that sense, the neutral point is the aerodynamic center of the airplane.*
- *From the stability condition*

$$\frac{dC_M}{d\alpha_w} = a_w(h_{CG} - h_{NP}) < 0$$

For a stable airplane

$$h_{CG} < h_{NP}$$

For larger magnitude of $|h_{CG} - h_{NP}|$ *, the airplane is more stable and less maneuverable.*

9.3.2 Static Margin

We define the static margin of an airplane as

$$static\ margin = h_{SM} = h_{NP} - h_{CG} \tag{9.20}$$

Typical static margin for transport airplanes is 0.05 to 0.1. General aviation airplanes have higher static margin, about 0.15 to 0.2, for higher stability (Raymer, 2006, pg. 473). In some fighter aircrafts, for "relaxed static stability" or greater maneuverability, the static margin is negative with computerized flight control (F16 and F22 have a static margin of -0.15).

From the equilibrium condition, we can determine the angle of attack at equilibrium

$$\circlearrowleft C_M = C_{M0} + \alpha_w a_w h_{SM} = 0$$

$$\alpha_{w,eq} = \frac{C_{M0}}{a_w h_{SM}} \tag{9.21}$$

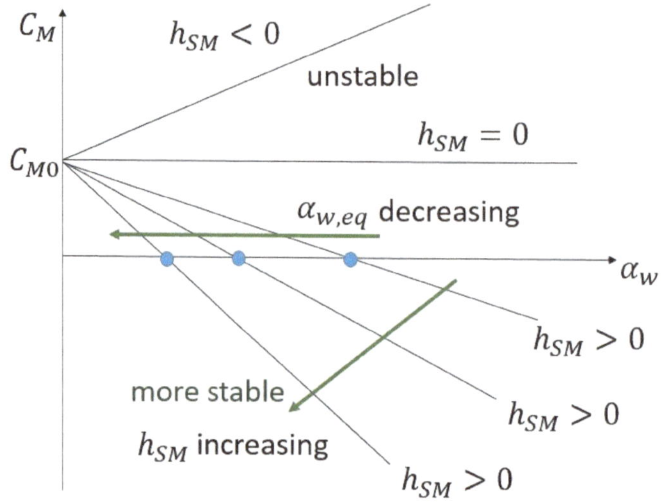

Figure 9.7: Stability and angle of attack

For larger static margin, the angle of attack at equilibrium is smaller.

Observation:

- In an airplane, with a given static margin, the pilot can change the equilibrium angle of attack by changing C_{M0}. The equilibrium angle of attack increases when pilot increases C_{M0}
- From Eqn. (9.14), the pilot can change C_{M0} by changing α_{t0} through elevator deflection.

9.3.3 Unconditionally Stable Airplanes

When the wing aerodynamic center is between the CG and the tail aerodynamic center, the CG is between the wing aerodynamic center and the nose of the airplane (see Figure 9.4). In this situation,

$$h_{CG} < 0$$

Therefore,

$$static\ margin = h_{NP} - h_{CG} = h_{NP} - (-|h_{CG}|) = h_{NP} + |h_{CG}| > 0$$

The airplane is unconditionally stable. In this arrangement, the tail produces negative lift and generally, tail airfoil has negative camber. The downside of this happy situation of unconditional stability is that it requires larger lift from the wing causing larger induced drag. We will revisit this situation in Example 9.7.

9.4 Trimmed Flight

9.4.1 Elevator and Equilibrium Angle of Attack

For a “trimmed flight,” the wing angle of attack is equal to the equilibrium angle of attack of Eqn. (9.21). The pilot can change the trimmed angle of attack by deflecting the elevator. A deflection of the elevator changes the lift curve of the tail and consequently the lift from the tail.

Assumption

- To simplify our calculations, we assume that an elevator deflection changes only the zero lift angle of attack of the tail but not the lift curve slope.

To analyze the effect of the elevator, we will re-visit Eqn. (9.14)

$$C_{M0} = C_{Macw} - a_w \alpha_{w0} h_{CG} + \eta V_H a_t (i_w - i_t - K a_w \alpha_{w0} + \alpha_{t0})$$

- When the elevator is up, $\alpha_{t0} > 0$ and C_{M0} increases
- When the elevator is down, $\alpha_{t0} < 0$ and C_{M0} decreases

From Eqn. (9.21), for the same static margin, a larger C_{M0} gives a larger wing angle of attack.

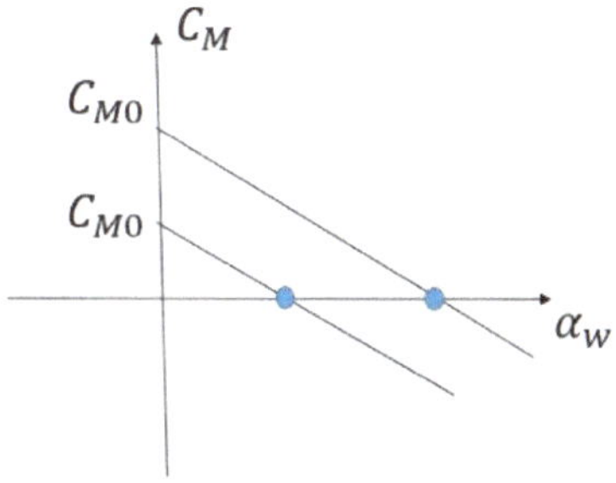

Figure 9.8: Angle of attack and C_{M0}

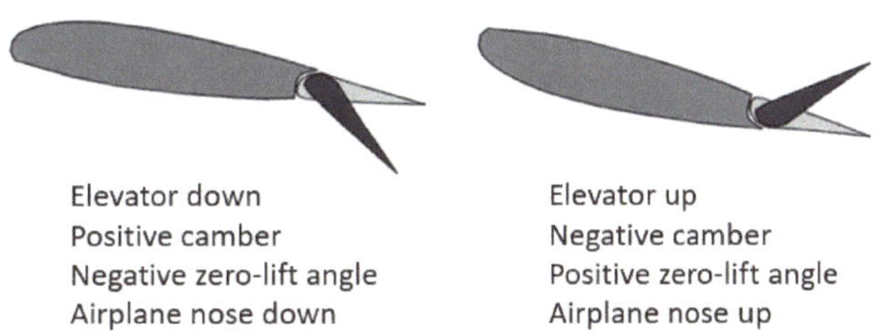

Figure 9.9: Elevator action

The yoke controls the elevator. When the pilot pulls the yoke, the elevator and the nose of the airplane go up.

Figure 9.10: Yoke

Observation:

- Airplane nose up or nose down does not mean that the airplane is climbing or descending. All of these calculations are for level flight.

Example-9.4

In an airplane, $a_w = 5.4$, $a_t = 4.8$, $\alpha_{w0} = -0.05$ rad, the angles of incidence of the wing and the tail are equal, wing $AR = 7.5$, $C_{Macw} = -0.02$, $h_{CG} = 0.2$, $V_H = 0.6$. Determine the trimmed angle of attack when (i) elevator is down with $\alpha_{t0} = -0.02$rad and (ii) elevator is up with $\alpha_{t0}=0.02$rad.

Efficiency factor

$$e = 1.78(1 - 0.045AR^{0.68}) - 0.64 = 0.8247$$

$$K = \frac{1}{\pi e AR} = 0.0515$$

From Eqn. (9.15)

$$C_{M\alpha} = -0.7917$$

Part (i): Tail airfoil has positive camber (elevator down)

$$\alpha_{t0} = -0.02 \ , \ \alpha_{w0} = -0.05$$

From Eqn. (9.14)

$$C_{M0} = 0.01817$$

$$\alpha_{w,eq} = -\frac{C_{M0}}{C_{M\alpha}} = 0.02296\, rad = 1.315^o$$

Part (ii): Tail airfoil has negative camber (elevator up)

$$\alpha_{t0} = 0.02 \ , \ \alpha_{w0} = -0.05$$

From Eqn. (9.14)

$$C_{M0} = 0.1219$$

$$\alpha_{w,eq} = -\frac{C_{M0}}{C_{M\alpha}} = 0.1539\, rad = 8.818^o$$

Elevator up decreases the lift from the tail and the nose of the airplane is up.

9.4.2 Lift in Trimmed Flight

The total lift from the wing and the tail is

$$L = L_w + L_t = \frac{1}{2}\rho V^2 S_w C_{Lw} + \frac{1}{2}\rho V_t^2 S_t C_{Lt} = \frac{1}{2}\rho V^2 S_w \left(C_{Lw} + \eta \frac{S_t}{S_w} C_{Lt}\right) \qquad (9.22)$$

The total lift coefficient of the airplane is

$$C_L = C_{Lw} + \eta \frac{S_t}{S_w} C_{Lt} \qquad (9.23)$$

From Eqn. (9.3)

$$C_{Lw,trim} = a_w(\alpha_{w,eq} - \alpha_{w0}) \qquad (9.24)$$

From Eqn. (9.10)

$$\begin{aligned} C_{Lt,trim} &= a_t(\alpha_{FRL} + i_t - \epsilon - \alpha_{t0}) \\ &= a_t(\alpha_{w,eq} - i_w + i_t - \epsilon - \alpha_{t0}) \end{aligned} \qquad (9.25)$$

9.4.3 Drag in Trimmed Flight

The tail lift coefficient is much smaller than the wing lift coefficient. Therefore, we ignore the induced drag from the tail in comparison with the induced drag from the wing.

$$C_{D,trim} = C_{D0} + K C_{Lw,trim}^2 \qquad (9.26)$$

9.4.4 Airspeed in Trimmed Flight

We can determine the airspeed at trimmed flight for an airplane of weight W as

$$W = \frac{1}{2}\rho V_{trim}^2 S_w C_{Lw,trim} \; ; \; V_{trim} = \sqrt{\frac{2W}{\rho S_w C_{Lw,trim}}} \qquad (9.27)$$

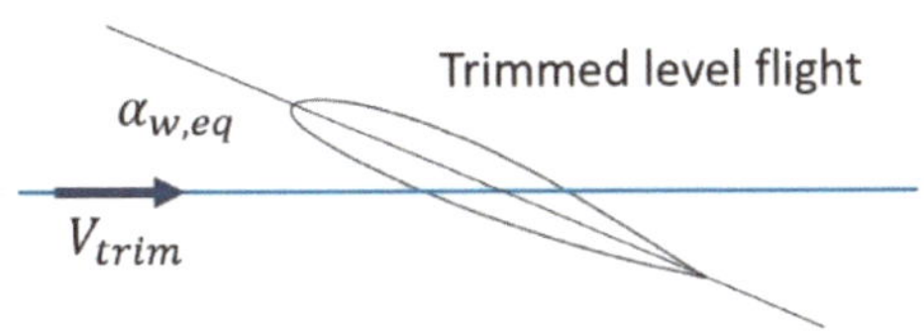

Figure 9.11: Trimmed level flight

We ignore the lift from the tail because it is much smaller than the lift from the wing.

9.4.4 Thrust in Trimmed Flight

From Eqns. (9.26) and (9.27),

$$T = D = \frac{1}{2}\rho V_{trim}^2 S_w C_{D,trim} \qquad (9.28)$$

Example 9.5

A 300,000lb airplane with wing area $S_w = 3000ft^2$, and wing aspect ratio of 7.5 is in level flight. The density of air is 0.0009 slug/ft^3. Other characteristics of the airplane are $C_{Macw} = -0.02$, $a_w = 5.8$, $\alpha_{w0} = -0.05rad$, $a_t = 5.1$, $i_w = i_t$, $V_H = 0.8$, $h_{CG} = 0.1$, $C_{D0} = 0.02$. Determine trimmed airspeed and drag when (i) the elevator is up with $\alpha_{t0} = 0.011rad$ and (ii) elevator is down with $\alpha_{t0} = -0.0044rad$

Part (i): This trimmed condition corresponds to elevator up.

From Eqn. (2.31),

$$e = 0.8247 \; ; \; K = 0.0515$$

From Eqns. (9.14), (9.15), and (9.16),

$$C_{M0} = 0.1042 \; ; \; C_{M\alpha} = -1.996 \; ; \; \alpha_{w,eq} = 0.0522rad$$

From Eqn. (9.24),

$$C_{Lw,trim} = 0.5928$$

From Eqn. (9.26),

$$C_{D,trim} = 0.03808$$

From Eqn. (9.27),

$$V_{trim} = \sqrt{\frac{2 \times 300000}{0.0009 \times 3000 \times 0.5928}} = 612.3\frac{ft}{s}$$

From Eqn. 9.28,

$$T = D = 19273lb$$

Part (ii)
This trimmed condition corresponds to elevator down.

From Eqns. (9.14), (9.15), and (9.16),

$$C_{M0} = 0.04764 \; ; \; C_{M\alpha} = -1.996 \; ; \; \alpha_{w,eq} = 0.02387rad$$

From Eqn. (9.24),

$$C_{Lw,trim} = 0.4284$$

From Eqn. (9.26),

$$C_{D,trim} = 0.02945$$

From Eqn. (9.27),

$$V_{trim} = \sqrt{\frac{2 \times 300000}{0.0009 \times 3000 \times 0.4284}} = 720.2 \frac{ft}{s}$$

From Eqn. 9.28,

$$T = D = 20619 lb$$

Observations:

- When the elevator is up, the airplane pitches up, the trimmed airspeed decreases, and drag force decreases. To decrease speed in level flight, decrease thrust and set elevator up.
- When the elevator is down, the airplane pitches down, the trimmed airspeed increases, and drag force increases. To increase speed in level flight, increase thrust and set elevator down.

9.4.5 Trimmed Cruise

A major portion of an airplane's mission is cruise, as it spends almost all of its flight time cruising. In this section, we will determine the incidence angles of the wing and the tail for the maximum lift to drag ratio, E_M. Instead of writing a system of algebraic equations, we will consider a practical example to demonstrate the process. Furthermore, we will proceed with a typical tail volume ratio of $V_H = 0.8$ and $h_{CG} = 0.2$. We will discuss the relationship between V_H and h_{CG} in Section 9.5. In an airplane with $K = 0.05$ and $C_{D0} = 0.02$, $C_{Macw} = -0.02$, $a_w = 5.7$, $a_t = 4.8$, $\alpha_{w0} = -0.05236 rad$, $\alpha_{t0} = 0$

$$C_{Lw,EM} = \sqrt{\frac{0.02}{0.05}} = 0.63$$

For the comfort of the passengers, we set

$$\alpha_{FRL} = 0 \ ; \ \alpha_w = \alpha_{FRL} + i_w = i_w$$

$$C_{Lw,EM} = 0.63 = a_w(i_w - \alpha_{w0}) = 5.7 \times (i_w + 0.05236) \ ; \ i_w = 0.05817$$

$$C_{Mw} = C_{Macw} + C_{Lw,EM} h_{CG} = 0.106$$

The downwash angle is

$$\epsilon = K C_{Lw,EM} = 0.0315$$

When the airfoil in the tail is symmetric and the elevator is neutral, the equilibrium equation is

$$C_{M,t} = -C_{M,w} = -0.106 = -0.9 \times 0.8 \times 4.8 \times (i_t - 0.0315) \ ; \ i_t = 0.06217$$

9.5 Tail Design

9.5.1 Tail Volume Coefficient and CG Location

So far, we solved the example problems by specifying a value for the tail volume coefficient. We will now estimate V_H by considering the landing requirement, especially during the flare (see Section 5.2). During landing, the pilot deploys the flaps and increases the angle of attack of the wing, to almost the stall angle of attack, to maximize the lift from the wing. Therefore, the moment from the wing is maximum during the flare and the tail volume coefficient plays a critical role in maintaining the equilibrium of the airplane.

When the pilot deploys the flap, we add two terms to C_{M0} of Eqn. (9.14). One term is an additional moment due to the additional lift from the flap.

$$\Delta C_{M,flap} = \Delta C_L h_{CG}$$

The other term is the additional downwash due to the additional lift from the flap.

$$\epsilon_{flap} = K\Delta C_L$$

Thus

$$C_{M0} = C_{Macw} + (\Delta C_L - a_w \alpha_{w0})h_{CG} + \eta V_H a_t(i_w - i_t - K a_w \alpha_{w0} + K\Delta C_L + \alpha_{t0}) \quad \text{(a)}$$

The flap also changes the zero-lift angle of attack of the wing (see Section 2.23).

We will estimate V_H and h_{CG} by satisfying three requirements.

- The designer sets the static margin to a positive value for stability.
- The airplane is in pitch equilibrium at an angle of attack set by the designer for maximum lift without stalling.

$$\alpha_w = \alpha_{w,eq} = \alpha_{stall}$$

- The pilot sets the zero-lift angle of attack of the tail at full downward deflection of the elevator for maximum lift from the tail to maintain equilibrium at maximum pitch up corresponding to stall.

We consider an airplane with the following properties: $K = 0.05$, $\Delta C_L = 0.4$, $a_w = 5.7$, $a_t = 5.1$, $C_{Macw} = -0.02$, $i_w = i_t$, $\alpha_{w0} = -0.1$.

The designer sets the following properties: $h_{SM} = 0.2$, $\alpha_w = 0.1 rad$.

The pilot fully deflects the elevator downward to attain $\alpha_{t0} = -0.07 rad$. From Eqn. (a),

$$C_{M0} = -0.02 + 0.97h_{CG} - 0.0987V_H$$

From Eqn. (9.21),

$$C_{M0} = \alpha_w a_w h_{SM} = 0.114$$

Thus, the condition for equilibrium is

$$0.97h_{CG} - 0.0987V_H = 0.134 \qquad \text{(b)}$$

The definition of h_{SM} is

$$h_{SM} = 0.2 = h_{NP} - h_{CG} = \eta V_H(1 - Ka_w)\frac{a_t}{a_w} - h_{CG}$$

By inserting the values of the known quantities

$$0.5419V_H - h_{CG} = 0.2 \qquad \text{(c)}$$

By solving Eqns. (b) and (c)

$$V_H = 0.77 \;\; ; \;\; h_{CG} = 0.22$$

9.5.2 Tail Size and Position

In this section, we will use $V_H = 0.85$ and $h_{CG} = 0.26$ to find the tail size and location. Typically, the aspect ratio of the tail is between three and five and the taper ratio of the tail is between 0.3 and 0.6 (see Raymer pg. 84).

We will use subscript *t* for the tail and *w* for the wing. From the definition of aspect ratio,

$$b_t^2 = AR_t S_t = AR_t b_t c_t \;\; ; \;\; b_t = AR_t c_t \;\; ; \;\; S_t = AR_t c_t^2$$

$$c_t = \frac{1}{2}\left(c_{R,t} + c_{T,t}\right) = \frac{\tau_t + 1}{2} c_{R,t}$$

$$S_t = AR_t \left(\frac{\tau_t + 1}{2}\right)^2 c_{R,t}^2$$

$$V_H = \frac{AR_t}{c_w S_w}\left(\frac{\tau_t + 1}{2}\right)^2 c_{R,t}^2 (x_t - x_{CG}) \qquad (9.31)$$

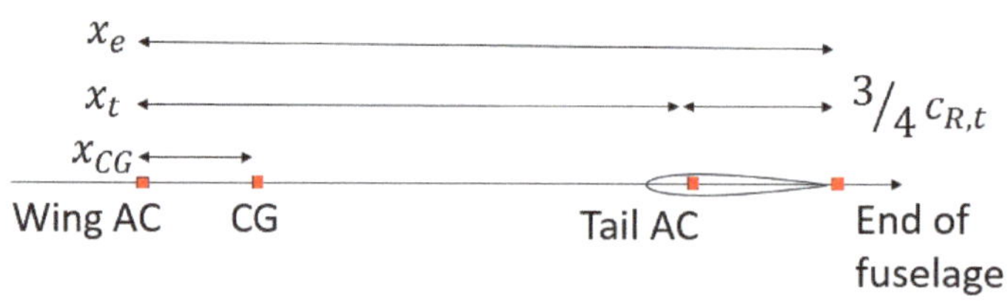

Figure 9.12: Attaching the tail

The tail must fit in the length between the aerodynamic center of the wing and the end of the fuselage. Therefore,

$$x_t = x_e - \frac{3}{4} c_{R,t} \tag{9.32}$$

By substituting Eqn. (9.32) in Eqn. (9.31)

$$g(c_{R,t}) = \frac{AR_t}{c_w S_w} \left(\frac{\tau_t + 1}{2}\right)^2 c_{R,t}^2 \left(x_e - \frac{3}{4} c_{R,t} - x_{CG}\right) - V_H = 0$$

We can find a graphical solution of the cubic equation for $c_{R,t}$ by plotting g versus the root chord of the tail.

We will use the following data: $S_w = 2000ft^2$, $c_w = 16.5ft$, $AR_t = 4$, $\tau_t = 0.45$, $x_e = 75ft$, $h_{CG} = 0.26$, $V_H = 0.85$.

$$x_{CG} = h_{CG} c_w = 4.29ft$$

With this data, we get the plot

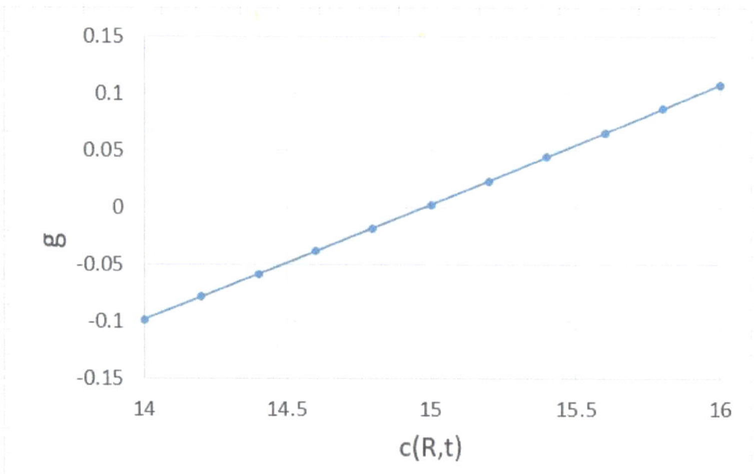

Figure 9.13: Solving cubic equation for the length of tail root-chord

$$c_{R,t} = 15ft \ ; \ \ c_{T,t} = \tau_t c_{R,t} = 6.75ft \ ; \ \ c_t = 0.5 \times \left(c_{R,t} + c_{T,t}\right) = 10.875ft$$

$$S_t = AR_t c_t^2 = 473ft^2 \ ; \quad x_t = x_e - \frac{3}{4}c_{R,t} = 63.75ft$$

The span of the tail is

$$b_t = \sqrt{AR_t S_t} = 43.5ft$$

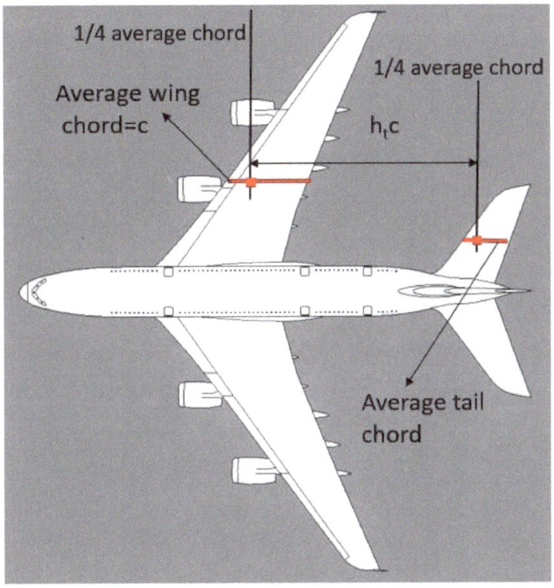

9.6 Spreadsheet

Stability calculations involve many formulas, are long, and are prone to mistakes when done manually. In Appendix F, we show the use of Excel spreadsheet for stability calculations.

10 Figures of Merit and Design

10.1 Figure of Merit

The definition of Figure of Merit in the Merriam-Webster Dictionary: a numerical quantity based on one or more characteristics of a system or device that represents a measure of efficiency or effectiveness. For an airplane, the performance characteristics we choose to measure the efficiency are fuel consumption, takeoff distance, fastest rate of climb, landing distance, and tightest turning radius.

In this chapter, we will describe the preliminary design process for a passenger carrying jet airplane. Such a design begins with three requirements: (i) the payload that includes number of passengers and their luggage, (ii) the range of the airplane, and (iii) the altitude and the length of the runway at the airports that the airplane will serve.

10.1.1 Cruise Characteristics

We will consider two cruise characteristics.

- The range for the flight program with constant airspeed and constant altitude (see Eqn. 3.32)

$$X_{V,\rho} = \frac{V}{c\sqrt{KC_{DO}}} tan^{-1}\left\{\frac{\left(\sqrt{KC_{DO}}\right)\zeta C_{L1}}{C_{D1} - \zeta K C_{L1}^2}\right\} \tag{10.1}$$

- Absolute ceiling is the altitude above which a level flight in equilibrium is impossible (see Eqn. 3.16)

$$\frac{\rho}{\rho_{SL}} = \frac{W}{E_M T_{SL}} \tag{10.2}$$

10.1.2 Takeoff Characteristics

We will consider two takeoff characteristics.

- The ground run (see Eqn. 4.11)

$$d = \frac{1.44\left({}^{W_{TO}}/_{S}\right)}{\rho g C_{L,max,TO}\left\{\left({}^{T}/_{W_{TO}}\right) - \mu\right\}} \tag{10.3}$$

- Fastest rate of climb (see Eqns. 4.25, 4.26, 4.27. 4.28)

$$Q_{FC} = \frac{\left({}^{T}/_{S}\right)}{6C_{DO}}\left[1 + \sqrt{1 + \frac{3}{\left\{E_M\left({}^{T}/_{W}\right)\right\}^2}}\right] \tag{10.4}$$

$$V_{FC} = \sqrt{\frac{2Q_{FC}}{\rho}} \tag{10.5}$$

$$D_{FC} = Q_{FC}SC_{D0} + K\frac{W^2}{Q_{FC}S} \tag{10.6}$$

$$sin\gamma_{FC} = \frac{(T - D_{FC})}{W} \ ; \ \ \dot{h}_{FC} = V_{FC}sin\gamma_{FC} \tag{10.7}$$

10.1.3 Landing Characteristics

We will consider one landing characteristic.

- The ground run (see Eqn. 5.15)

$$s_g = \frac{1.323\left({}^{W_{LA}}/_{S}\right)}{\rho g C_{L,max,LA}\left\{\left({}^{T_R}/_{W_{LA}}\right) + \mu\right\}} \tag{10.8}$$

10.1.4 Turning Flight Characteristics

We will consider one characteristic for turning flight.

- The relevant equations for calculating the radius of the tightest turn are (see Eqns. 6.21, 6.22, 6.23, 6.25, 6.26)

$$Q_{TT} = 2K\frac{\left({}^{W}/_{S}\right)}{\left({}^{T}/_{W}\right)} \tag{10.9}$$

$$V_{TT} = \sqrt{\frac{2Q_{TT}}{\rho}} \tag{10.10}$$

$$n_{TT} = \sqrt{2\left[1 - \left\{\frac{2KC_{D0}}{\left({}^{T}/_{W}\right)^2}\right\}\right]} \tag{10.11}$$

$$r_{TT} = \frac{V_{TT}^2}{g}\frac{1}{\sqrt{n_{TT}^2 - 1}} \tag{10.12}$$

$$C_{L,TT} = \frac{n_{TT}\left(\frac{W}{S}\right)}{Q_{TT}} = \frac{n_{TT}}{2K}\left(\frac{T}{W}\right) \leq C_{L,max} \tag{10.13}$$

10.2 Requirements

We will do a preliminary design of an airplane with the following requirements:

- Carries *120* passengers with each passenger carrying *75lb* of luggage
- Carries *2* pilots and *5* cabin crews with each crew member carrying *75lb* of luggage
- The airplane has a range of *2500miles*
- The airplane can serve an airport at an altitude of *5000ft*
- The absolute ceiling of the airplane is *35000ft*

10.3 Lift Coefficient

We choose the aspect ratio of our wing to be *7* and calculate the induced drag coefficient from Eqn. (2.31).

$$e = 1.78(1 - 0.045AR^{0.68}) - 0.64 = 0.8392 \ ; \ K = \frac{1}{\pi eAR} = 0.05419$$

We choose NACA 2412 as the airfoil in our wing. From Figure 2.38, the properties of this airfoil are

$$a_0 = 0.1 \ per \ degree \ ; \ \alpha_{L=0} = -2^o \ ; \ \alpha_{stall} = 8^o$$

From Eqn. (2.32)

$$a = \frac{0.1}{1 + 0.05419 \times 0.1} = 0.09946 \ per \ degree$$

$$C_{L,Max} = 0.09946 \times (\alpha_{stall} - \alpha_{L=0}) = 0.9946$$

The wing has Fowler flaps, the ratio of the flapped area and the wing area is *0.40*, and the ratio of chords with and without flap is *1.2*. We can determine the maximum lift coefficient for takeoff by following the method of Example 2.19. Assume that the angle of attack at takeoff is the stall angle of attack of 8°.

$$\Delta C_{l,Max} = 1.3 \times 1.2 = 1.56 \ ; \ \Delta C_{L,Max} = 0.9 \times 1.56 \times 0.4 = 0.5616$$

$$\Delta C_{L,Max,TO} = 0.7 \times 0.5616 = 0.3931$$

$$\Delta \alpha_{L=0} = -10 \times 0.4 = -4^o \ ; \ \alpha_{L=0} = -2 - 4 = -6^o$$

$$C_L = 0.09946 \times (8 + 6) = 1.392$$

$$C_{L,Max,TO} = C_L + \Delta C_{L,Max,TO} = 1.392 + 0.3931 = 1.785$$

Similarly, we can determine the maximum lift coefficient for landing by following the method of Example 2.20. Assume that the angle of attack at landing is the stall angle of attack of 8°.

$$\Delta C_{l,Max} = 1.3 \times 1.2 = 1.56 \ ; \ \Delta C_{L,Max} = 0.9 \times 1.56 \times 0.4 = 0.5616$$

$$\Delta C_{L,Max,LA} = 1 \times 0.2808 = 0.5616$$

$$\Delta \alpha_{L=0} = -15 \times 0.4 = -6^o \ ; \ \alpha_{L=0} = -2 - 6 = -8^o$$

$$C_L = 0.09946 \times (8 + 8) = 1.591$$

$$C_{L,Max,LA} = C_L + \Delta C_{L,Max,LA} = 1.591 + 0.5616 = 2.153$$

10.4 First Estimate

The design is an iterative process. In the first estimate, we include only the weight of the fuselage and the payload. To find the weight of the fuselage, we decide the seating arrangement and find the dimensions *a, b, c,* and *d* of Figure A.1 in Appendix A.

10.4.1 Fuselage

The figure below shows the seating arrangement.

- The business class has *20* seats in *5* rows.
- The pitch in the business class is *65in*
- The economy class has *16* rows with *6* seats and *1* exit row with *4* seats.
- The pitch in the economy class is *36in*
- The width of the aisle is *32in* in business class and *27in* in economy class.
- The height of the aisle is *80in*
- Radius of the fuselage is *7.1ft*

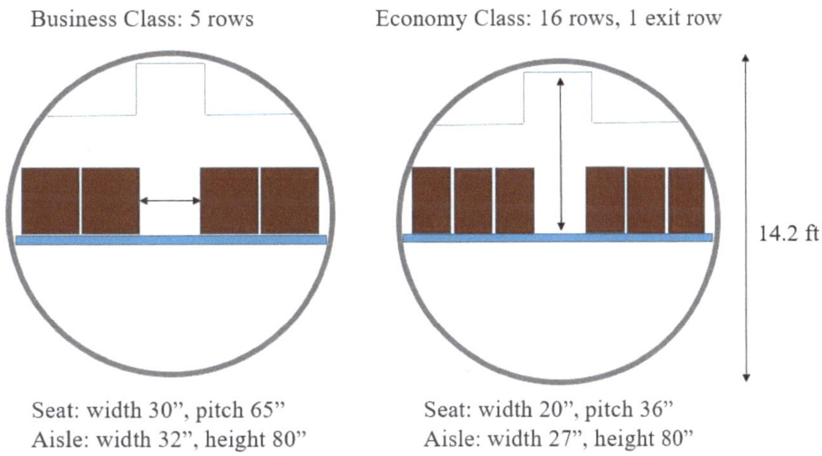

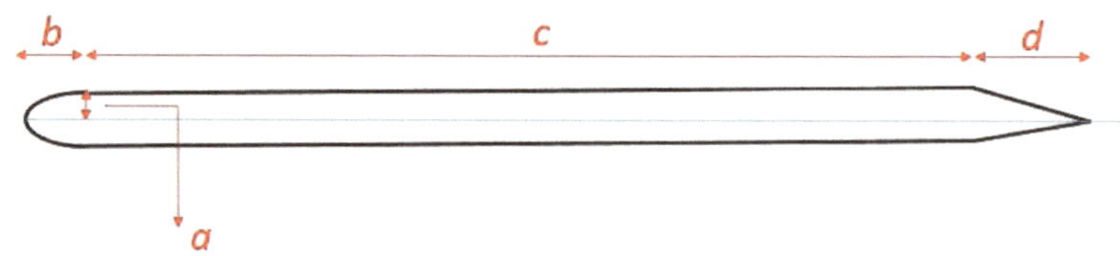

1	**Cockpit (half ellipsoid)**	
2	b (ft)	14
3	**Cylinder**	
4	Door (ft)	4
5	2 Lavatories, galley, storage (ft)	6
6	5 Rows, 65in pitch, business seating (ft)	27.08
7	17 Rows, 36in pitch, economy seating (ft)	51
8	4 Lavatories (ft)	6
9	Galley (ft)	6
10	c (ft)	100.08
11	**Cone**	
12	d (ft)	28
13	**Radius**	
14	a (ft)	7.1

We calculate the surface area of the fuselage by using the formulas from Appendix A.

Areas (sq. ft.)	
Half Ellipsoid	353.9
Cylinder	4465
Cone	644.3
Total	5463

From Table 15.2 (Raymer 2006, pg. 455), the approximate empty weights of components are

Component	**Transports (lb/sq.ft)**
Wing	10
Horizontal Tail	5.5
Vertical Tail	5.5
Fuselage	5

Figure 10.1: Approximate weight of components

Thus, the weight of the fuselage is

$$W_{fuselage} = 5463 \times 5 = 27314lb$$

We will ignore the weight of the facilities such as, lavatories, galley, etc.

10.4.2 Passenger and Luggage

We have *127* seats that weigh *30lb* per seat

$$W_{misc} = 127 \times 30 = 3810lb$$

We have *127* persons on board with average weight of *165lb* per person and each carrying luggage of *75lb*

$$W_{pass} = 127 \times (165 + 75) = 30480lb$$

Hence,

$$W_1 = W_{fuselage} + W_{misc} + W_{pass} = 61604lb$$

10.4.3 Wing Weight

Let the wing planform area be *S*. From Figure 10.1, weight of the wing is

$$W_{wing} = 10S$$

$$W_2 = W_{wing} + W_1 = 61604 + 10S$$

We will design two models of our airplane: (i) Model-A with wing loading of $90lb/ft^2$ and (ii) Model-B with wing loading of $100lb/ft^2$. For Model-A,

$$\frac{61604 + 10S}{S} = 90 \ ; \quad S = 770\,ft^2 \ ; \ W_2 = 69305lb$$

For Model-B

$$\frac{61604 + 10S}{S} = 100 \ ; \quad S = 684\,ft^2 \ ; \ W_2 = 68449lb$$

10.4.4 Parasite Drag

The calculation of the parasite drag coefficient begins with the skin friction coefficient. From Section 2.13

$$C_f = 0.00275$$

From Figure-2.51, for thickness to chord ratio of 0.12 for NACA 2412, the form factor *F=1.2*. From Eqn. (2.34)

$$C_{D0,wing} = 0.00275 \times 1.2 \times \frac{S_{wet}}{S_{ref}} = 0.00275 \times 1.2 \times 2 = 0.0066$$

For the fuselage, from Eqn. (2.35)

$$f = \frac{\ell}{\sqrt{(4/\pi)A_{max}}} = \frac{14 + 100.8 + 28}{14.2} = 10.01 \;;\; F = 1.077$$

From Eqn. (2.34),

Model	F	S_{wet}	S_{ref}	$C_{D0,fuselage}$
A	1.077	5463	770	0.021
B	1.084	5463	684	0.024

Hence, the total parasite drag coefficient is

Model	$C_{D0,wing}$	$C_{D0,fuselage}$	C_{D0}
A	0.0066	0.021	0.0276
B	0.0066	0.024	0.0302

10.4.5 Engine

We calculate E_M from Eqn. (2.41) and calculate the thrust at sea level from Eqn. (10.2). Density at the ceiling of *35000ft* is $0.000738 slug/ft^3$ and density at sea level is $0.00238 slug/ft^3$.

Model	E_M	W_2 (lb)	T_{SL} (lb)
A	12.93	69305	17283
B	12.35	68449	17862

Engine data for civil turbofan engines are available at the URL: http://www.jet-engine.net/civtfspec.html.

We chose two GE-CF34-8C1 engines.

- Each engine produces *13790lb* thrust at sea level
- Each engine weighs *2350lb*
- The specific fuel consumption is *0.37*

10.4.5 Fuel Weight

For cruise at the altitude of *30000ft* ($\rho = 0.0008907 slug/ft^3$), the available thrust is (see Eqn. 8.19)

$$T_{30000} = \frac{\rho_{30000}}{\rho_{SL}} T_{SL} = \frac{0.0008907}{0.00238} \times 27580 = 10321 lb$$

From Eqn. (3.7), we can calculate the thrust required for a given airspeed. The figures below show the variation of required thrust with airspeed. The thrust versus airspeed plot is U-shaped (see Figure 3.2) and we have shown only the right branch of the plot. At *70%* of the available thrust, *7225lb*, the airspeed for both the models is about *800ft/s* (*545mph*).

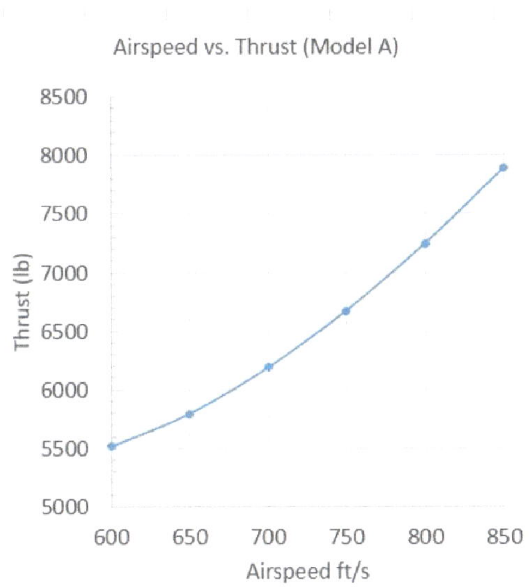

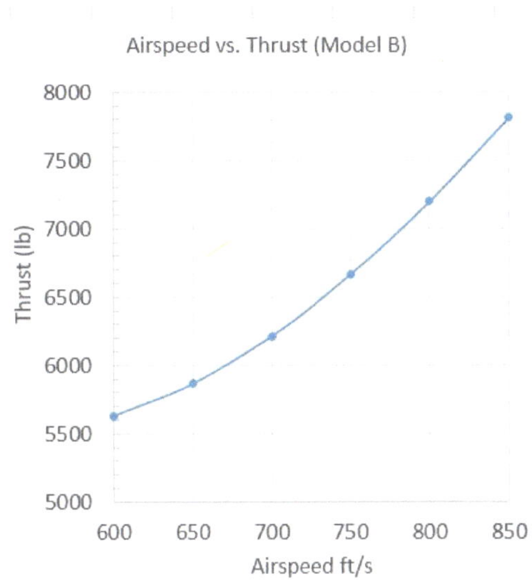

For an airspeed of *800ft/s* (*545mph*), at an altitude of *30000ft* ($\rho = 0.0008907 slug/ft^3$), for a distance of *2500miles,* we can calculate the fuel weight ratio from Eqn. (10.1). We find the lift and drag coefficients from

$$C_{L1} = \frac{2W_2}{\rho V^2 S} \; ; \; C_{D1} = C_{D0} + KC_{L1}^2$$

We take the specific fuel consumption as *0.37lb/h.lb* for the chosen engine, and we find the fuel consumption for Models-A and B from

$$Fuel\, Weight = W_{jet} = \zeta W_2$$

Model	W_2 (lb)	K	C_{D0}	C_{L1}	C_{D1}	ζ	W_{jet} (lb)
A	69304	0.05419	0.0276	0.3158	0.0330	0.1725	11955
B	68449	0.05419	0.0302	0.3508	0.0369	0.1730	11844

10.5 Second Estimate

We update the weight by including the fuel weight and the engine weight.

$$W_3 = W_1 + W_{jet} + W_{engine}$$

Model	W_3 (lb)
A	78260
B	78148

10.5.1 Wing Weight and Geometry

For Model-A,

$$\frac{78260 + 10S}{S} = 90 \ ; \ \ S = 978.2\ ft^2$$

For Model-B

$$\frac{78148 + 10S}{S} = 100 \ ; \ \ S = 868.3\ ft^2$$

For takeoff, we include all the weights. For landing, we exclude the fuel weight W_{jet}. Thus,

$$W_{TO} = W_3 + W_{wing} \ ; \ \ W_{LA} = W_3 + W_{wing} - W_{jet}$$

Model	W_{TO} (lb)	W_{LA} (lb)
A	88042	76087
B	86832	74987

From Eqn. (2.30),

Model	S (ft²)	AR	b (ft)	c (ft)
A	978.2	7	82.75	11.82
B	868.3	7	77.96	11.14

10.5.2 Parasite Drag

From Eqn. (2.34),

Model	F	S_{wet}	S_{ref}	$C_{D0,fuselage}$
A	1.077	5463	978.2	0.0165
B	1.084	5463	868.3	0.0186

Hence, the total parasite drag coefficient is

Model	$C_{D0,wing}$	$C_{D0,fuselage}$	C_{D0}
A	0.0066	0.0165	0.0231
B	0.0066	0.0186	0.0252

10.6 Performance

We will now compare the figures of merit for Model-A and Model-B.

10.6.1 Cruise Airspeed and Thrust

For cruise at the altitude of *30000ft* ($\rho = 0.0008907 slug/ft^3$), the available thrust is (see Eqn. 8.19)

$$T_{30000} = \frac{\rho_{30000}}{\rho_{SL}} T_{SL} = \frac{0.0008907}{0.00238} \times 27580 = 10321lb$$

From Eqn. (3.7), we can calculate the thrust required for a given airspeed. The figures below show the variation of required thrust with airspeed. The thrust versus airspeed plot is U-shaped (see Figure 3.2) and we have shown only the right branch of the plot. At *70%* of the available thrust, *7225lb*, the airspeed for both the models is about *730ft/s* (*497mph*). At this airspeed, the flight time for *2500miles* is *5.02hours*.

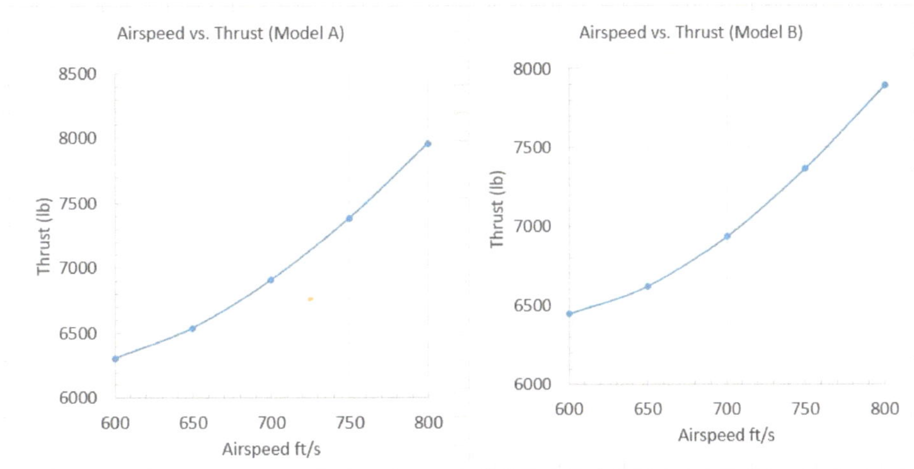

10.6.2 Fuel Consumption

For an airspeed of *730ft/s* (*497mph*), at an altitude of *30000ft* ($\rho = 0.0008907 slug/ft^3$), for a distance of *2500miles,* we can calculate the fuel weight ratio from Eqn. (10.1). We find the lift and drag coefficients from

$$C_{L1} = \frac{2W_{TO}}{\rho V^2 S} \ ; \ C_{D1} = C_{D0} + KC_{L1}^2$$

We take the specific fuel consumption as *0.37lb/h.lb* for the chosen engine, and we find the fuel consumption for Models-A and B from

$$Fuel\ Weight = \zeta W_{TO}$$

Model	W_{TO} (lb)	K	C_{D0}	C_{L1}	C_{D1}	ζ	Fuel (lb)
A	88042	0.05419	0.0231	0.3792	0.0309	0.1462	12870
B	86832	0.05419	0.0252	0.4214	0.0348	0.1477	12822

The fuel consumption with the Second Estimate differs by about 8% from the fuel consumption with the First Estimate. Therefore, we decide to not continue with a Third Estimate.

10.6.3 Takeoff Distance at 5000ft

We will now compare the takeoff distances for Models-A and B at an airport at an altitude of *5000ft* ($\rho = 0.002048\ slug/ft^3$). We correct the available thrust for altitude.

$$T_{5000} = \frac{0.002048}{0.00238} \times 27580 = 23733lb$$

We assume that the takeoff thrust is *90%* of available thrust.

$$T_{TO} = 21359lb$$

We take the friction coefficient to be *0.03* and we have found the maximum lift coefficient at takeoff in Section 10.3. We find the takeoff distance from Eqn. (10.6).

Model	W_{TO} (lb)	S (ft^2)	$C_{L,Max,TO}$	T (lb)	d (ft)
A	88042	978.2	1.785	21359	5179
B	86832	868.3	1.785	21359	5664

Model-B has a larger wing loading, compared to Model-A, thus, its takeoff distance is longer.

10.6.4 Fastest Rate of Climb at 5000ft

We will now compare the fastest rate of climb for Models-A and B at an airport at an altitude of *5000ft* ($\rho = 0.002048\ slug/ft^3$). We correct the available thrust for altitude and assume that the climb thrust is *80%* of available thrust (see Section 10.6.3)

$$T = 18986lb$$

We fine E_M by using Eqn. (2.41) and find the fastest rate of climb by using the Eqns. (10.4) to (10.7).

Model	E_M	W_{TO} (lb)	S (ft^2)	T (lb)	Q_{FC} (lb/ft^2)	V_{FC} (ft/s)	D_{FC} (lb)	Rate of Climb (ft/s)
A	14.12	88042	978.2	18986	469.8	677.3	11545	57.24
B	13.52	86832	868.3	18986	487.5	690.0	11644	58.34

10.6.5 Landing Distance at 5000ft

We will now compare the landing distances for Models-A and B at an airport at an altitude of *5000ft* ($\rho = 0.002048\ slug/ft^3$). We correct the available thrust for altitude and assume that the reverse thrust is *40%* of available thrust (see Section 5.4).

$$T_{5000} = \frac{0.002048}{0.00238} \times 27580 = 23733lb$$

$$T_R = 0.4 \times 23773 = 9509lb$$

We take the friction coefficient to be *0.03* and we have found the maximum lift coefficient at landing in Section 10.3. From Eqn. (10.8),

Model	W_{LA} (lb)	S (ft^2)	$C_{L,Max,LA}$	T_R (lb)	s (ft)
A	76087	978.2	2.153	9509	4677
B	74987	868.3	2.153	9509	5132

Model-B has a larger wing loading, compared to Model-A, thus, its landing distance is longer.

10.6.6 Tightest Turning Radius

For a turning flight, the load factor and airspeed combination must lie within the flight envelop (see Section 2.24). Thus, it is essential to know the bounds of the load factor to avoid stall. To establish the bounds, we choose a set of load factors $n > 1$ (see Example 6.6). For each load factor, we calculate the thrust to weight ratio from Eqn. (10.9) and the lift coefficient from Eqn. (10.13). We show the variation of lift coefficient with load factor for Models-A and B in the figure below.

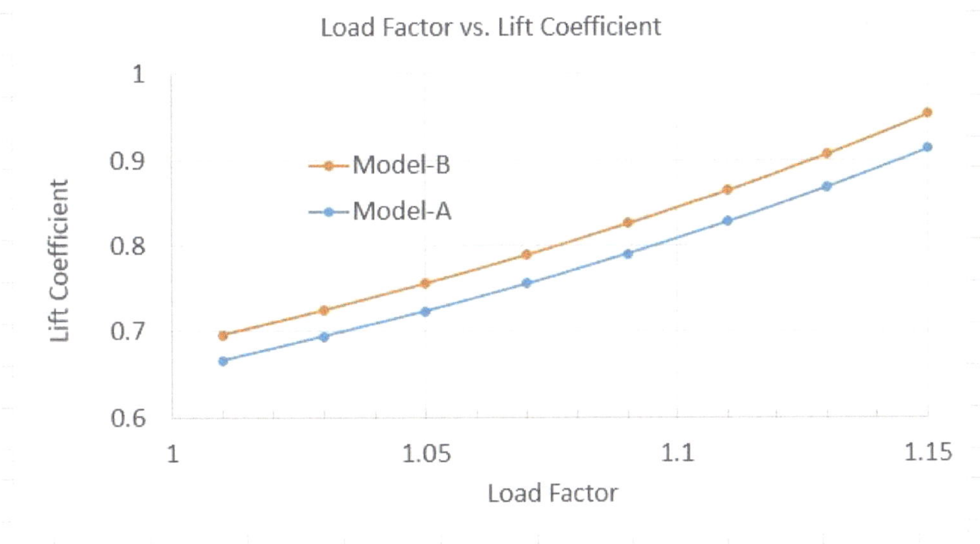

The maximum lift coefficient for the wing is *0.9946* (see Section 10.3). We choose the maximum, allowed load factor to be *1.11*. This load factor corresponds to the banking angle of

$$\beta_{TT} = cos^{-1}\left(\frac{1}{n_{TT}}\right) = 25.7^o$$

Model	n_{TT}	T/W	$C_{L,TT}$	V_{TT} (ft/s)	r (ft)
A	1.11	0.08603	0.8276	308.7	5210
B	1.11	0.08984	0.8643	318.4	5543

10.7 Wing Geometry

The formula in Eqn. (2.32) corrects the lift curve slope of an airfoil for one geometric characteristic of the wing, the aspect ratio. In Figure 2.55, we have shown that this correction is adequate for preliminary design. After the preliminary design, we make improvements and add detail to our design by considering other geometric characteristics of the wing, such as, taper, twist, sweepback, and dihedral. For this purpose, we use elaborate numerical techniques.

Here, we will use the eight-vortex lifting line model (see Section 2.20.2) to explore the effect of taper ratio on the lift coefficient. For a wing with NACA 2412 airfoil, aspect ratio of *7*, area of $978ft^2$, and at an angle of attack of 4°, we find that the wing with the taper ratio of *0.3* has the largest lift coefficient.

Taper Ratio	Span (ft)	Root Chord (ft)	Tip Chord (ft)	Mean Chord (ft)	C_L
0.6	82.74	14.78	8.865	11.82	0.4669
0.5	82.74	15.76	7.88	11.82	0.4686
0.4	82.74	16.89	6.754	11.82	0.4698
0.3	82.74	18.18	3.94	11.82	0.4700
0.2	82.74	19.7	2.94	11.82	0.4684
0.1	82.74	21.49	2.149	11.82	0.4635

10.8 Tail Geometry and Location

We will use $V_H = 0.8$, $h_{CG} = 0.2$, tail aspect ratio of *4*, and tail taper ratio of *0.45* to find the tail size and location (see Section 9.5.2). From Eqns. (9.17) and (9.20), and by assuming that the wing and tail have the same lift curve slope, the neutral point and the static margin are

$$h_{NP} = \eta V_H (1 - K a_w) \frac{a_t}{a_w} = 0.9 \times 0.8 \times (1 - 0.05419 \times 5.73) = 0.4964$$

$$h_{SM} = h_{NP} - h_{CG} = 0.2964$$

We will use subscripts *t* and *w* for the tail and the wing. From the definition of aspect ratio and Eqn. (9.31)

$$V_H = \frac{AR_t}{c_w S_w} \left(\frac{\tau_t + 1}{2} \right)^2 c_{R,t}^2 (x_t - x_{CG})$$

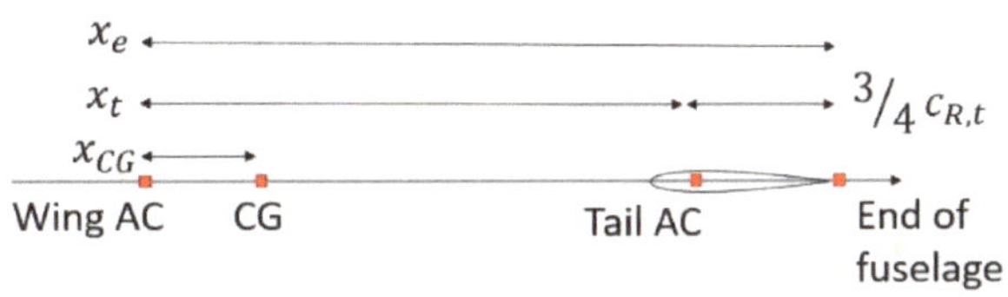

The tail must fit in the length between the aerodynamic center of the wing and the end of the fuselage. Therefore,

$$x_t = x_e - \frac{3}{4} c_{R,t} \tag{9.32}$$

We can find a graphical solution of the following cubic equation for the root chord of the tail

$$g(c_{R,t}) = \frac{AR_t}{c_w S_w}\left(\frac{\tau_t + 1}{2}\right)^2 c_{R,t}^2 \left(x_e - \frac{3}{4} c_{R,t} - x_{CG}\right) - V_H = 0$$

We plot the function $g(c_{R,t})$ with the following data: $S_w = 980ft^2$, $c_w = 11.82ft$, $AR_t = 4$, $\tau_t = 0.45$, $h_{CG} = 0.2$, $V_H = 0.8$.

$$x_{CG} = h_{CG} c_w = 2.36ft$$

Assuming that the location of the center of gravity is at the midpoint of the fuselage

$$x_e = 2.36 + 0.5(length\ of\ fuselage) = 2.36 + 71.4 = 73.76\ ft$$

With this data, we get the plot

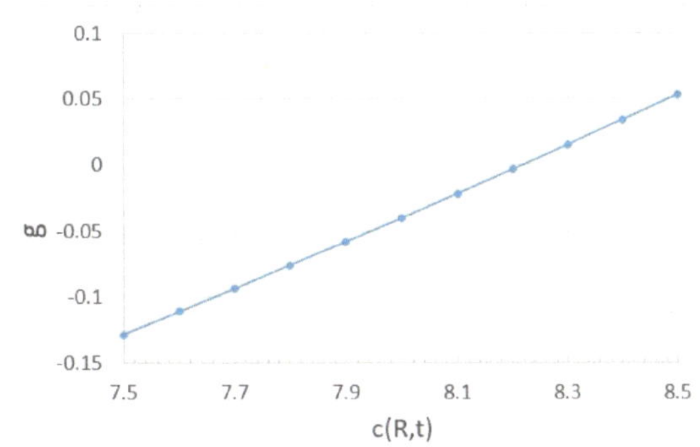

$$c_{R,t} = 8.2ft\ ;\ \ c_{T,t} = \tau_t c_{R,t} = 3.69ft\ ;\ \ c_t = 0.5 \times (c_{R,t} + c_{T,t}) = 5.945ft$$

$$S_t = AR_t c_t^2 = 141.4ft^2\ ;\ \ x_t = x_e - \frac{3}{4} c_{R,t} = 67.61ft$$

The span of the tail is

$$b_t = \sqrt{AR_t S_t} = 23.78ft$$

Appendix A

Surface Area of a Fuselage

We need the surface area of the fuselage to calculate the parasite drag of the fuselage (see Section 2.20.8). To estimate the surface area, we partition the fuselage into three pieces – a semi-ellipsoid, a cylinder, and a cone.

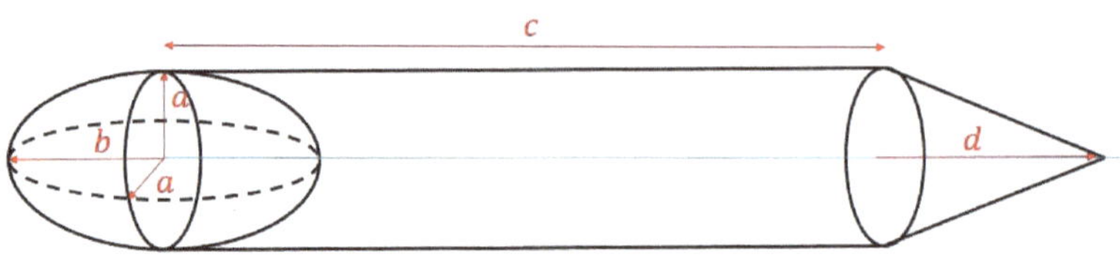

Figure A.1: Fuselage

The surface area of the semi-ellipsoid is

$$S_{ell} = 2\pi \left\{ \frac{2(ab)^{1.6} + (a^2)^{1.6}}{3} \right\}^{1/1.6}$$

The surface area of the cylinder is

$$S_{cyl} = 2\pi ac$$

The surface area of the cone is

$$S_{cone} = \pi a\sqrt{d^2 + a^2}$$

Appendix B

Excel Solver

Solver is an "Add-On" utility in Microsoft Excel for solving complex equations, minimization, and maximization of functions. We have to install Solver in Excel before we want to use it for the first time.

To add Solver:

1. File Menu > Options > Add-Ins > Manage Excel Add-ins > Go

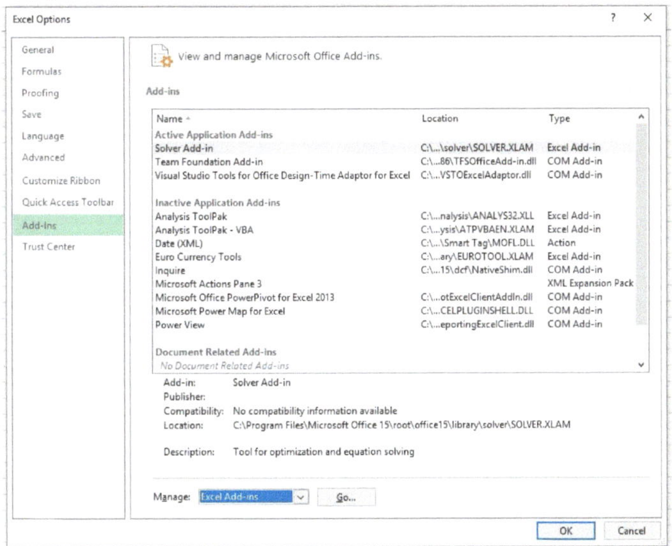

2. Check Solver Add-In > OK

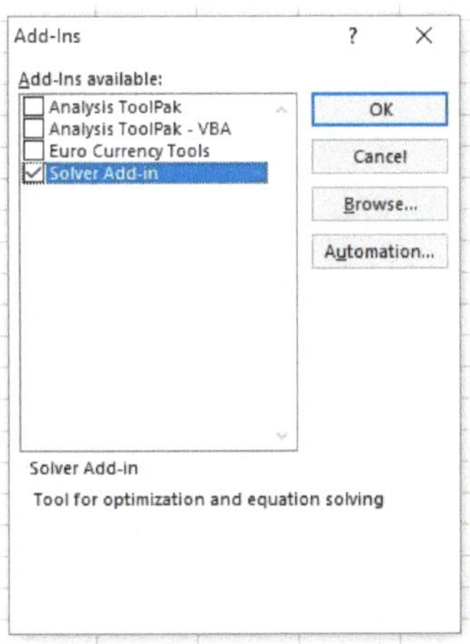

We want to solve the equation

$$f(x) = 24(27 + x)^x - 35 = 0$$

In an Excel worksheet, the value of x is in cell *B1*. We entered the function *f(x)* in cell *B2*.

	A	B
1	x	
2	f(x)	=24*(27+B1)^B1-35

Select the DATA tab at the top and click “Solver”

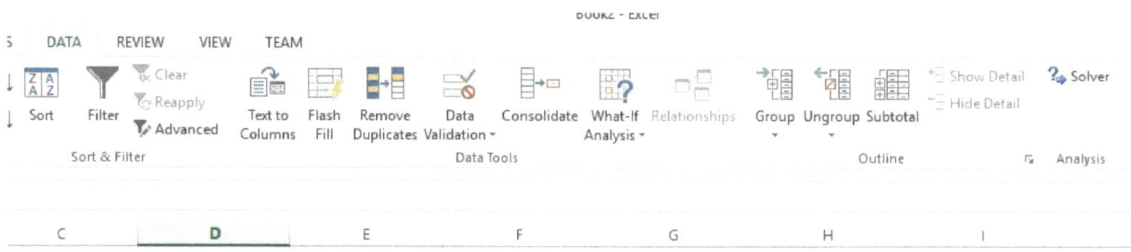

You will get the “Solver Window”

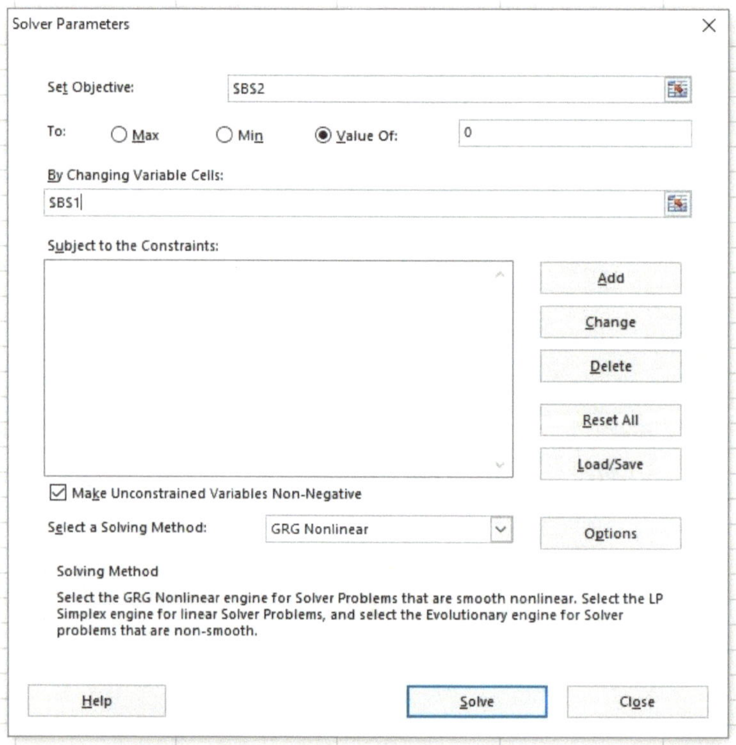

- Set Objective B2
- Value of 0
- By Changing Variable Cells B1
- Click Solve

	A	B
1	x	0.114329519953133
2	f(x)	=24*(27+B1)^B1-35

Appendix C

Fuel Consumption in V and ρ Constant Flight Program

<u>*Example C.1*</u>

A *250000lb* airplane has wing area $S = 2400ft^2$, $C_{D0} = 0.02$, induced drag coefficient $K = 0.05$, thrust specific fuel consumption $c = 0.6lb/h.lb$. At an altitude of *20000ft,* determine the "cruise fuel weigh ratio" for a range of *4000miles* at a speed of *400miles/h.*
(ρ *at 20000ft =0.001267slug/ft*^{3})

We enter the input data in cells *B1* through *B7.* We calculate velocity in *ft/s* in cell *B8*; enter Eqn. (3.31) in cells *B10* and *B11,* and enter Eqn. (3.32) in cell *B12.*

	A	B
1	Weight	250000
2	Wing Area (S)	2400
3	Density (rho)	0.001267
4	Parasite Drag (CD0)	0.02
5	Induced Drag Factor (K)	0.05
6	Specific Fuel Consumption (c)	0.6
7	Velocity (mph)	400
8	Velocity (ft/s)	=B7*5280/3600
9	zeta	0.348717511863948
10	CL1	=B1/(0.5*B3*B8*B8*B2)
11	CD1	=B4+B5*B10*B10
12	X	=(B7/(SQRT(B5*B4)))*ATAN((SQRT(B5*B4)*B9*B10)/(B11-B9*B5*B10*B10))/B6

Set Solver window as shown below and click Solve.

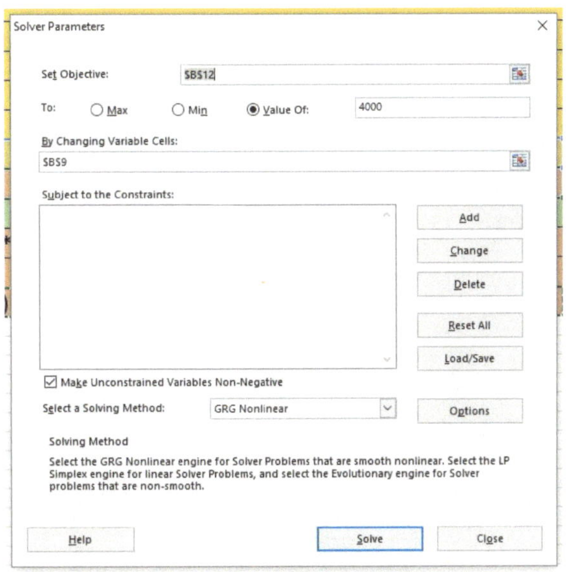

Solver displays solution in cell *B9* as $\zeta = 0.3487$.

Appendix D

Climb

Example D.1

An airplane has a thrust to weight ratio of 0.25 at sea level, wing loading of 110lb/ft^2, C_{D0} = 0.03, induced drag coefficient K = 0.06. Determine the velocity when the climb angle at sea level is *6°*.

We enter the input data in cells *B1* through *B5.* We calculate velocity in *ft/s* in cell *B6* and in mph in cell *C6,* calculate dynamic pressure in cell *B7*; enter Eqn. (4.18) in cell *B8,* calculate climb angle in radians in cell *B9*, and in degrees in cell *C9*.

	A	B	C
1	W/S	110	
2	T/W	0.25	
3	CD0	0.03	
4	K	0.06	
5	rho (SL)	0.002377	
6	V	682.347313743563	=B6*3600/5280
7	Q	=0.5*B5*B6^2	
8	sin(gamma)	=B2-((B7*B3/B1)+(B4*B1/B7))	
9	Gamma	=ASIN(B8)	=B9*180/PI()

Set Solver window as shown below and click Solve.

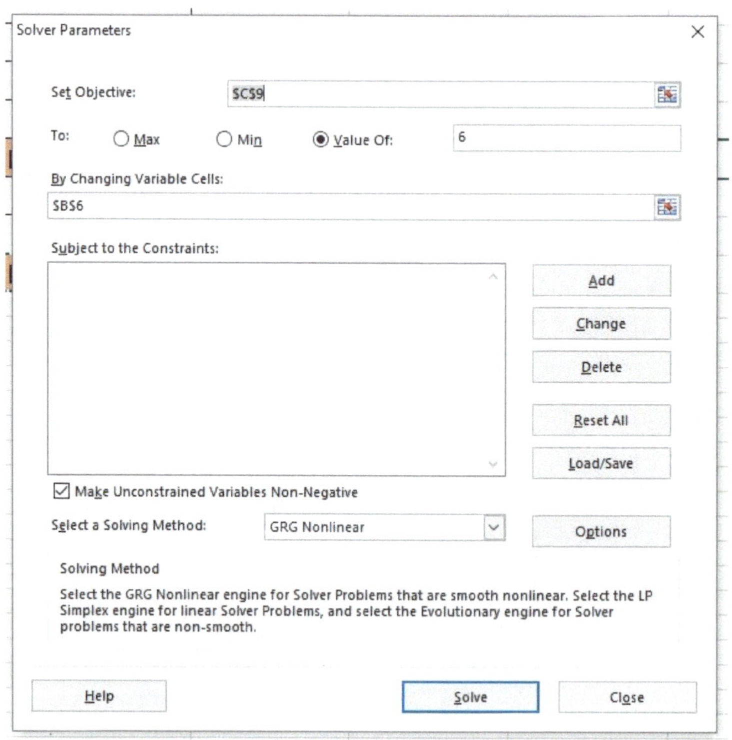

Solver displays solution in cell *B6* as $V = 682.3\frac{ft}{s}$.

Example D.2

A 200000lb airplane has a wing area of 2000ft^2, $C_{D0} = 0.02$, $K = 0.05$, thrust to weight ratio at sea level of 0.31. Determine the velocity for fastest rate of climb at sea level.

We enter the input data in cells *B1* through *B5*. We calculate velocity in *ft/s* in cell *B6* and in mph in cell *C6*, calculate dynamic pressure in cell *B7*; enter Eqn. (4.18) in cell *B8*, enter Eqn. (4.20) in cell *B9*.

	A	B	C
1	W/S	100	
2	T/W	0.31	
3	CD0	0.02	
4	K	0.05	
5	rho (SL)	0.002377	
6	V	669.248268335472	=B6*3600/5280
7	Q	=0.5*B5*B6^2	
8	sin(gamma)	=B2-((B7*B3/B1)+(B4*B1/B7))	
9	Rate of Climb	=B6*B8	

Set Solver window as shown below to maximize cell *B9*, and click Solve.

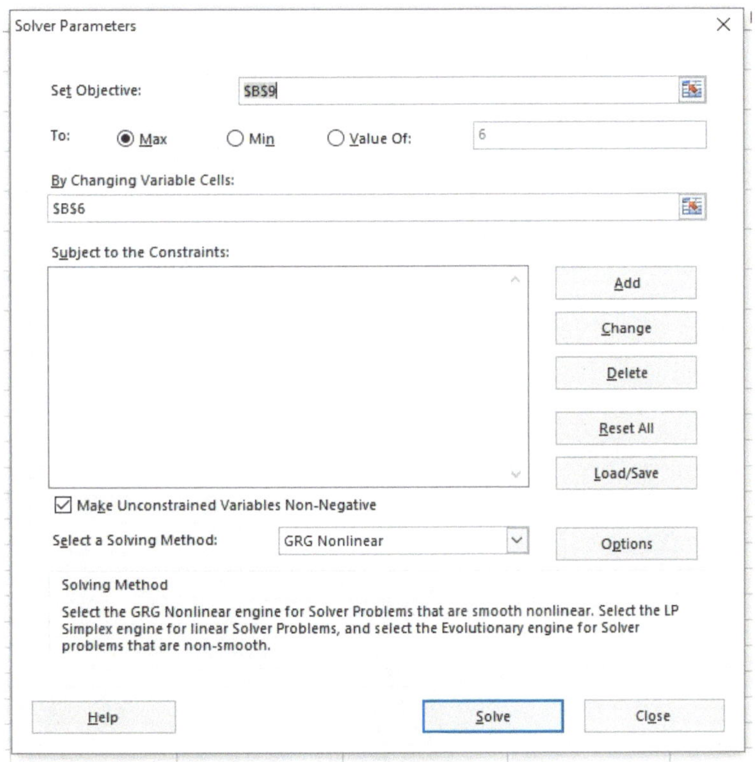

Solver displays solution in cell *B6* as $V = 669.2\frac{ft}{s}$. Compare this solution with the solution of Example 4.5.

Appendix E

Tightest Turn

Example E.1

An airplane has a wing loading of $100lb/ft^2$, $C_{D0} = 0.02$, $K = 0.05$. Determine the radius of the tightest turn at sea level. The maximum lift coefficient with flaps is *1.2*. The maximum thrust to weight ratio is *0.25* at sea level.

We enter the input data in cells *B1* through *B5*. Enter Eqn. (6.21) in cell *B6*, Eqn. (6.22) in cell *B7*, Eqn. (6.23) in cell *B8*, Eqn. (6.26) in cell *B9*, Eqn. (6.25) in cell *B10*, Eqn. (6.24) in cell *B11*, and convert bank angle from radians to degrees in cell *C11*.

	A	B	C
1	W/S	100	
2	T/W	0.06	
3	CD0	0.015	
4	K	0.05	
5	rho	0.002377	
6	Q(TT)	=2*B4*B1/B2	
7	V(TT)	=SQRT(2*B6/B5)	
8	n(TT)	=SQRT(2*(1-(2*B4*B3/(B2^2))))	
9	CL(TT)	=B8*B1/B6	
10	r(TT)	=B7^2/(32.2*SQRT(B8^2-1))	
11	beta	=ACOS(1/B8)	=B11*180/PI
12			

Thrust to weight ratio of *0.05* is insufficient because load factor is less than one.

	A	B	C
1	W/S	100	
2	T/W	0.05	
3	CD0	0.015	
4	K	0.05	
5	rho	0.002377	
6	Q(TT)	200	
7	V(TT)	410.2186535	
8	n(TT)	0.894427191	
9	CL(TT)	0.447213595	
10	r(TT)	#NUM!	
11	beta	#NUM!	####
12			

We show two solutions for thrust to weight ratios of *0.06* and *0.08*.

	A	B	C
1	W/S	100	
2	T/W	0.06	
3	CD0	0.015	
4	K	0.05	
5	rho	0.002377	
6	Q(TT)	166.6666667	
7	V(TT)	374.4766834	
8	n(TT)	1.08012345	
9	CL(TT)	0.64807407	
10	r(TT)	10667.66373	
11	beta	0.387596687	22.2
12			

	A	B	C
1	W/S	100	
2	T/W	0.08	
3	CD0	0.015	
4	K	0.05	
5	rho	0.002377	
6	Q(TT)	125	
7	V(TT)	324.306321	
8	n(TT)	1.237436867	
9	CL(TT)	0.989949494	
10	r(TT)	4481.315117	
11	beta	0.629839534	36.1

It is counter-intuitive to observe that velocity decreases as the thrust to weight ratio increases.

Observations

- In a higher velocity turn, the dynamic pressure is higher.
- For higher dynamic pressure, the lift coefficient is smaller.
- For smaller lift coefficient, the induced drag is smaller.
- Smaller induced drag requires smaller thrust.
- For higher velocity, the load factor is smaller, radius of tightest turn is larger, and the bank angle is smaller.
- Higher load factor and higher bank angle make the passengers uncomfortable

A turn is not possible at thrust to weight ratio of *0.1* because required lift coefficient is more than the maximum lift coefficient and the airplane stalls.

	A	B	C
1	W/S	100	
2	T/W	0.1	
3	CD0	0.015	
4	K	0.05	
5	rho	0.002377	
6	Q(TT)	100	
7	V(TT)	290.0683917	
8	n(TT)	1.303840481	
9	CL(TT)	1.303840481	
10	r(TT)	3123.172142	
11	beta	0.696698361	39.9

To study the effect of wing loading on the tightest turn, we keep the thrust to weight ratio fixed and change the wing loading.

	A	B	C
1	W/S	100	
2	T/W	0.07	
3	CD0	0.015	
4	K	0.05	
5	rho	0.002377	
6	Q(TT)	142.8571429	
7	V(TT)	346.6980404	
8	n(TT)	1.178030179	
9	CL(TT)	0.824621125	
10	r(TT)	5994.709493	
11	beta	0.556943564	31.9

	A	B	C
1	W/S	90	
2	T/W	0.07	
3	CD0	0.015	
4	K	0.05	
5	rho	0.002377	
6	Q(TT)	128.5714286	
7	V(TT)	328.9066404	
8	n(TT)	1.178030179	
9	CL(TT)	0.824621125	
10	r(TT)	5395.238543	
11	beta	0.556943564	31.9

Observations

- Wing loading has no effect on the load factor, lift coefficient, and bank angle.
- Velocity for the tightest turn decreases as the wing loading decreases.
- Radius of the tightest turn decreases as the wing loading decreases.

Appendix F

Static Stability

- We enter the input data in cells *B1* through *B7* and *B9* through *B14*.

	A	B
1	**Static Stability**	
2	AR(w)	7.5
3	C(Macw)	-0.02
4	a(w)	5.7
5	alpha(w0)	-0.03
6	i(w)	0.04
7	a(t)	5.1
8	alpha(t0)	0
9	i(t)	0.02
10	VH	0.8
11	h(CG)	0.1
12	Weight	300000
13	Density	0.001
14	S(W)	3000

- We insert the formulas in rows *B15* through *B24* and *B26* through *B29.*
- We insert the Solver variables in *B8* and *B25*.

15	e	=1.78*(1-0.045*B2^0.68)-0.64
16	K	=1/(PI()*B15*B2)
17	C(M0)	=B3-B4*B5*B11+0.9*B10*B7*(B6-B9-B16*B4*B5+B8)
18	C(M_alpha)	=B4*B11-0.9*B10*B7*(1-B16*B4)
19	alpha(w)	=-1*B17/B18
20	C(Lw)	=B4*(B19-B5)
21	C(Lt)	=B7*(B19-B6+B9-B16*B20-B8)
22	CMw	=B3+B20*B11
23	CMt	=-0.9*B10*B21
24	h(NP)	=0.9*B10*(1-B16*B4)*B7/B4
25	V	=SQRT(2*B12/(B13*B14*B20))
26	CD0	0.02
27	C(D)	=B26+B16*B20*B20
28	D	=0.5*B13*B25^2*B14*B27
29	E	=B20/B27

Trimmed Level Flight

Example F.1

For an airplane, $a_w = 5.7$, $a_t = 5.1$, $\alpha_{w0} = -0.03rad$, $i_w = 0.04rad$, $i_t = 0.02rad$, wing $AR = 7.5$, $C_{Macw} = -0.02$, $h_{CG} = 0.1$, $V_H = 0.8$. The *300,000 lb* airplane with wing area of $3000ft^2$ is in level flight at an altitude with density $0.001 slug/ft^3$. Determine α_{t0} to maintain (i) $V_{trim} = 650\ ft/s$ and (ii) $V_{trim} = 700\ ft/s$. Also, determine the corresponding $\alpha_{w,eq}$ and drag force.

8	alpha(t0)	0.0012432	8	alpha(t0)	-0.005066
9	i(t)	0.02	9	i(t)	0.02
10	VH	0.8	10	VH	0.8
11	h(CG)	0.10000	11	h(CG)	0.10000
12	Weight	300000	12	Weight	300000
13	Density	0.001	13	Density	0.001
14	S(W)	3000	14	S(W)	3000
15	e	0.8247361	15	e	0.8247361
16	K	0.0515	16	K	0.0515
17	C(M0)	0.10742	17	C(M0)	0.08425
18	C(M_alpha)	-2.02491	18	C(M_alpha)	-2.02491
19	alpha(w)	0.05305	19	alpha(w)	0.04161
20	C(Lw)	0.473374	20	C(Lw)	0.408163
21	C(Lt)	0.037969	21	C(Lt)	0.028911
22	CMw	0.027337	22	CMw	0.020816
23	CMt	-0.027337	23	CMt	-0.020816
24	h(NP)	0.45525	24	h(NP)	0.45525
25	V	650.00	25	V	700.00
26	CD0	0.02	26	CD0	0.02
27	C(D)	0.03153	27	C(D)	0.02857
28	D	19983	28	D	21001
29	E	15.012771	29	E	14.284831

(i) Set Objective *B25* to Value of *650* by Changing Variable Cells *B8*; Click Solve; Check Keep Solver Solution; Click OK.

$$\alpha_{t0} = 0.001243rad\ ;\ \alpha_{w.eq} = 0.05305rad\ ; D = 19983lb$$

(ii) Set Objective *B25* to Value of *700* by Changing Variable Cells *B8*; Click Solve; Check Keep Solver Solution; Click OK.

$$\alpha_{t0} = -0.005066rad\ ;\ \alpha_{w,eq} = 0.04161rad\ ; D = 21001lb$$

Observations:

- Lower V_{trim} requires up elevator, larger C_{M0} in cell *B17*, larger angle of attack, and smaller thrust.
- Higher V_{trim} requires down elevator, smaller C_{M0} in cell *B17*, smaller angle of attack, and larger thrust.

Trimmed and Steady Ascent

During an ascent, the flight path makes angle with the horizon. For steady flight, the equilibrium equations for the airplane are (see Section 4.2)

$$T - D - W sin\gamma = 0$$
$$L - W cos\gamma = 0$$

Thus, for ascent, thrust is larger than drag and lift is approximately equal to weight for small flight path angle. The pilot enters a straight descent by increasing the angle of attack and simultaneously increasing the engine thrust.

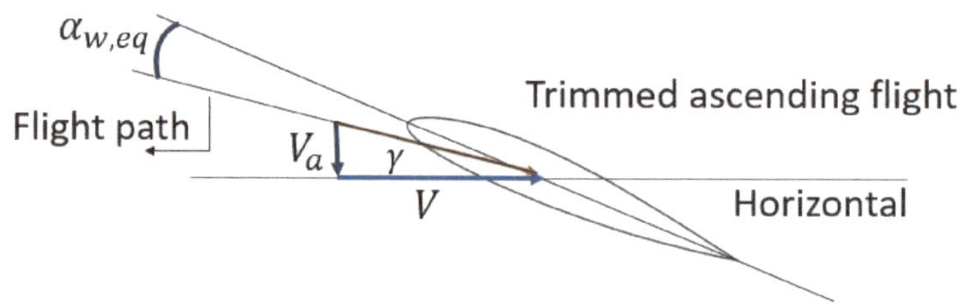

Example F.2

For an airplane, $a_w = 5.7$, $a_t = 5.1$, $\alpha_{w0} = -0.03 rad$, $i_w = 0.04 rad$, $i_t = 0.02 rad$, wing $AR = 7.5$, $C_{Macw} = -0.02$, $h_{CG} = 0.1$, $V_H = 0.8$. The airplane has a forward velocity of 650 ft/s and an ascending velocity of 50 ft/s. The weight of the airplane is 300,000 lb and density is 0.001 slug/ft^3. Determine $\alpha_{w,eq}$ and thrust.

$$V_a = 50 ^{ft}/_{s} \; ; V = 650 ^{ft}/_{s} \; ; \; V_{trim} = \sqrt{650^2 + 50^2} = 651.9 \frac{ft}{s} \; ; \; \gamma = 0.07677\, rad$$

Set Objective *\$B\$25* to Value of *651.9* by Changing Variable Cells *\$B\$8*; Click Solve; Check Keep Solver Solution; Click OK.

$$\alpha_{w,eq} = 0.05256\, rad$$

The angle between the wing and the horizontal is

$$\alpha_{w,eq} + \gamma = 0.1293\, rad$$

The angle between the FRL and the horizontal is

$$\alpha_{w,eq} + \gamma - i_w = 0.1293 - .04 = 0.0893 rad = 5.118^0$$

$$T = D + Wsin\gamma = 20015 + 300000 \times 0.0767 = 43025lb$$

8	alpha(t0)	0.0009766
9	i(t)	0.02
10	VH	0.8
11	h(CG)	0.10000
12	Weight	300000
13	Density	0.001
14	S(W)	3000
15	e	0.8247361
16	K	0.0515
17	C(M0)	0.10644
18	C(M_alpha)	-2.02491
19	alpha(w)	0.05256
20	C(Lw)	0.470618
21	C(Lt)	0.037586
22	CMw	0.027062
23	CMt	-0.027062
24	h(NP)	0.45525
25	V	651.90
26	CD0	0.02
27	C(D)	0.03140
28	D	20015
29	E	14.989011

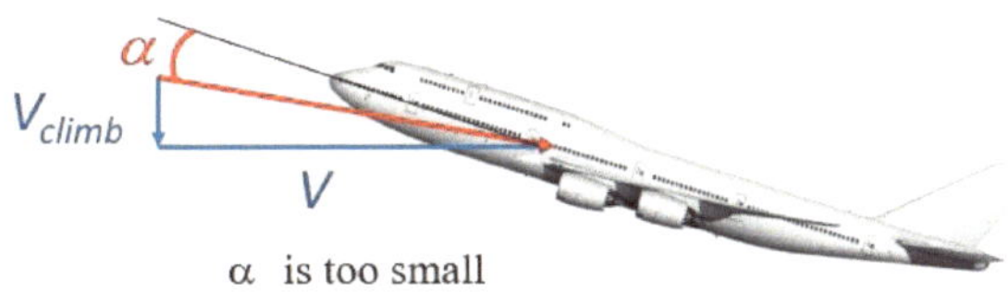

α is too small
$L < W$

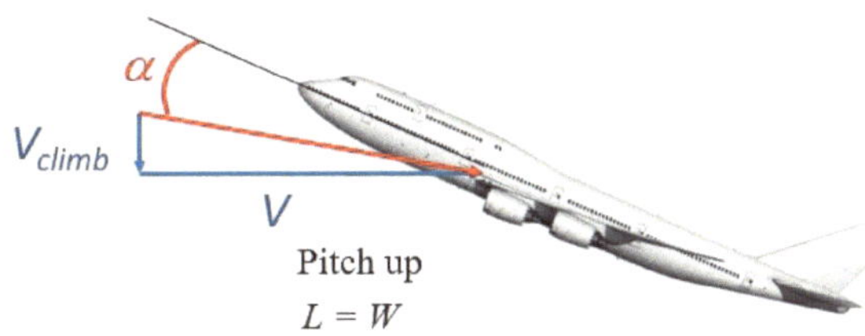

Pitch up
$L = W$

Trimmed and Steady Descent

During a descent, the flight path makes angle with the horizon. For steady flight, the equilibrium equations for the airplane are (see Section 5.2)

$$T - D + Wsin\gamma = 0$$
$$L - Wcos\gamma = 0$$

Thus, for descent, thrust is smaller than drag and lift is approximately equal to weight, for small flight path angle. The pilot enters a straight descent by increasing the angle of attack and simultaneously decreasing the engine thrust.

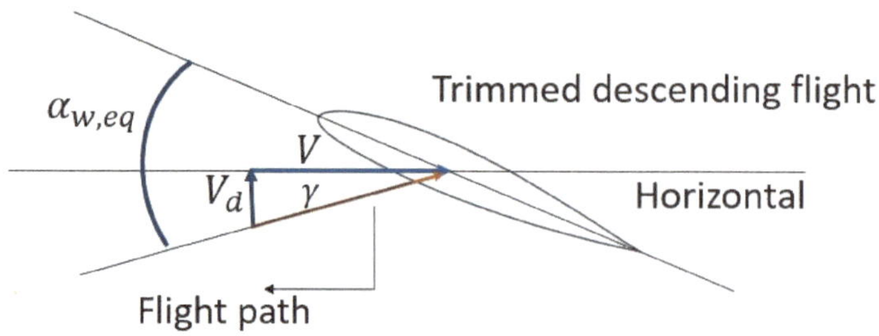

Example F.3

For an airplane, $a_w = 5.7$, $a_t = 5.1$, $\alpha_{w0} = -0.03rad$, $i_w = 0.04rad$, $i_t = 0.02rad$, wing $AR = 7.5$, $C_{Macw} = -0.02$, $h_{CG} = 0.1$, $V_H = 0.8$. The airplane has a forward velocity of 650 ft/s and descending velocity of 50 ft/s. The weight of the airplane is 300,000 lb and density is 0.002 slug/ft^3. Determine $\alpha_{w,eq}$ and thrust.

$$V_d = 50\frac{ft}{s}\ ;\ V = 650\frac{ft}{s}\ ;\ V_{trim} = \sqrt{650^2 + 20^2} = 651.9\frac{ft}{s}\ ;\ \gamma = 0.07677\ rad$$

Set Objective *\$B\$25* to Value of *651.9* by Changing Variable Cells *\$B\$8*; Click Solve; Check Keep Solver Solution; Click OK.

$$\alpha_{w,eq} = 0.01128\ rad$$

The angle between the wing and the horizontal is

$$\alpha_{w,eq} - \gamma = -0.06549rad$$

The angle between the FRL and the horizontal is

$$\alpha_{w,eq} + \gamma - i_w = -0.06528 - .04 = -0.1055rad = -6.044^0$$

$$T = D - Wsin\gamma = 29131 - 300000 \times 0.0767 = 6121lb$$

8	alpha(t0)	-0.021788
9	i(t)	0.02
10	VH	0.8
11	h(CG)	0.10000
12	Weight	300000
13	Density	0.002
14	S(W)	3000
15	e	0.8247361
16	K	0.0515
17	C(M0)	0.02285
18	C(M_alpha)	-2.02491
19	alpha(w)	0.01128
20	C(Lw)	0.235308
21	C(Lt)	0.004904
22	CMw	0.003531
23	CMt	-0.003531
24	h(NP)	0.45525
25	V	651.90
26	CD0	0.02
27	C(D)	0.02285
28	D	29131
29	E	10.298247

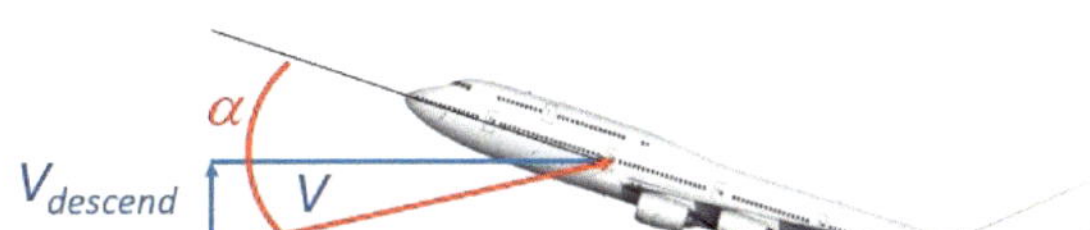

α is too large
$L > W$

Pitch down
$L = W$

Appendix G

Standard Atmosphere

Altitude (1000 ft)	Temperature (Rankine)	Pressure (lb/sq.ft)	Density (slug/cu.ft.)
0	518.7	2116.2	0.0023769
1	515.1	2040.9	0.0023081
2	511.5	1967.7	0.0022409
3	508.0	1896.7	0.0021752
4	504.4	1827.7	0.0021109
5	500.8	1760.9	0.0020482
6	497.3	1696.0	0.0019869
7	493.7	1633.1	0.0019270
8	490.2	1572.1	0.0018685
9	486.6	1512.9	0.0018113
10	483.0	1455.6	0.0017555
11	479.5	1400.1	0.0017011
12	475.9	1346.2	0.0016480
13	472.3	1294.1	0.0015961
14	468.8	1243.6	0.0015455
15	465.2	1194.8	0.0014962
16	461.7	1147.5	0.0014480
17	458.1	1101.7	0.0014011
18	454.5	1057.5	0.0013553
19	451.0	1014.7	0.0013107
20	447.4	973.3	0.0012673
21	443.9	933.3	0.0012249
22	440.3	894.6	0.0011836
23	436.7	857.2	0.0011435
24	433.2	821.2	0.0011043
25	429.6	786.3	0.0010663
26	426.1	752.7	0.0010292
27	422.5	720.3	0.0009931
28	419.0	689.0	0.0009580
29	415.4	658.8	0.0009239
30	411.8	629.7	0.0008907
31	408.3	601.6	0.0008584
32	404.7	574.6	0.0008270
33	401.2	548.5	0.0007966
34	397.6	523.5	0.0007670
35	394.1	499.3	0.0007382

Bibliography

1. Abbott, I.H., von Doenhoff, A.E., Stivers, L.S., Summary of Airfoil Data, National Advisory Committee for Aeronautics, Report 824, 1945.

2. Anderson, J.D., Fundamental of Aerodynamics, 6th Ed., McGraw-Hill, New York, 2017.

3. Biermann, D. and Hartman, E.P., Tests of Five Full-Scale Propellers in the Presence of a Radial and a Liquid-cooled Engine Nacelle, Including Tests of Two Spinners, National Advisory Committee for Aeronautics, Report 642, 1937.

4. Cavallo, B., Subsonic Drag Estimation Methods, US Naval Air Development Center, Report NADC-AW-6604, 1966.

5. Hall, A.W., A Statistical Study of Normal Load Factor Just Prior to Ground Contact for Five Light Airplanes, NASA TN D-5767, 1970.

6. Raymer, D.P., Aircraft Design: A Conceptual Approach Fourth Edition, AIAA Education Series, J.A. Schetz Senior Editor-in-Chief, 2006.

7. Schlichting, H., Boundary Layer Theory, McGraw-Hill, New York, 1979.

8. Wetmore, J.W., The Rolling Friction of Several Airplane Wheels and Tires and the Effect of Rolling Friction on Takeoff, National Advisory Committee for Aeronautics, Report 583, 1937.

CPSIA information can be obtained
at www.ICGtesting.com
Printed in the USA
LVHW072144140819
627724LV00009B/125/P

* 9 7 8 0 9 9 9 7 4 6 6 1 5 *